PRACTICAL FARMING

WORKING DOGS
Training for Sheep and Cattle

Colin Seis

INKATA PRESS

INKATA PRESS

A DIVISION OF BUTTERWORTH-HEINEMANN

AUSTRALIA

BUTTERWORTH-HEINEMANN
North Tower 1–5 Railway Street
Chatswood NSW 2067

BUTTERWORTH-HEINEMANN
18 Salmon Street
Port Melbourne 3207

UNITED KINGDOM
BUTTERWORTH-HEINEMANN LTD
Oxford

USA
BUTTERWORTH-HEINEMANN
Stoneham

National Library of Australia Cataloguing-in-Publication entry

Seis, Colin.
Working dogs: training for sheep and cattle.

Includes index.
ISBN 0 7506 8920 X.

1. Sheep dogs - Training. 2. Cattle dogs - Training.
I. Title. (Series: Practical farming).

636.737

©1995 Colin Seis

Typeset by Ian MacArthur, Hornsby Heights, NSW.
Printed in Australia by Ligare Pty Ltd, Riverwood, NSW

4

Contents

Introduction

A well bred and well trained working dog is an invaluable asset to every farmer and stockman.

The process of training a pup to be a dependable working dog is in itself a rewarding task. This book covers the basic obedience training and follows on to detailed paddock and yard training as the dog matures. By the age of twelve months the dog — Kelpie, Border Collie or Australian Cattle Dog — should be fully broken in and able to handle stock with confidence and efficiency.

The terms used will be familiar to some of you and may be totally foreign to others. Common obedience commands like **sit**, **stay** and **come** form the basic essentials of control. Getting a pup used to being on a chain, training it to stay on the back of a truck and initial introduction to stock, form an integral and important progression in the dog's training. Understanding balance, teaching a dog to cover, cast, obey direction commands, jump and control barking are a further progression of your dog's training to become a useful worker. Most well bred working dogs are more than capable of learning these tasks, many of their actions are instinctive and only need refining by a kind and patient handler.

Acknowledgement

The author and the publisher wish to thank the following for their assistance in the preparation of this book:

Mrs B. Cooper
Marion Wilson
Steve Bilson
Daryl Cluff
Les Jones
Mark Powell
Peter Pryde (Pryde's Tucker Bags Pty Ltd)
The Australian Cattle Dog Society of NSW
The Working Kelpie Council of Australia Inc
Winona Merino & Kelpie Studs

1

History of the Australian Working Dog

1

Since earliest recorded history, dogs have helped to gather, yard and handle stock. Early Egyptian drawings show shepherds and their dogs with grazing sheep near the pyramids. In Roman times dogs were already classified into groups according to the work they did. In England one of the oldest publications *Englishe Dogges*, written by John Keys in 1576 classified dogs of the day into various groups. *Canis Pastoralis* or Shepherd Dogge was one of these groups.

Over the centuries dogs have been progressively bred to become more and more specialised for specific tasks. The list is endless — guard dogs, seeing-eye dogs for the blind, pig dogs and of course the sheep and cattle dogs bred to have the physical attributes required for the arduous work they do.

Border Collie

From 'History of the Border Collie' by Stephen Bilson in *Stories From All Around Australia*, ABC Books.

In September 1893 a Working Colley pup was born near the border of England and Scotland that would change the way sheepdogs worked around the world! This was the birth of the father of all modern Border Collies, an exceptional dog now referred to as Old Hemp.

The full history of the Border Collie can however really go back much further than this to the Roman invasion of Britain in 55 BC. The Romans brought with them their own breed of sheepdog. These were described as large and heavy boned and carried either a long or a short coat and were usually black with tan and white markings. These dogs spent most of their time fending off wolves, foxes and guarding the shepherd's possessions and sheep from other humans. It is thought that these dogs resembled the large Bernese Mountain Dog that is still popular today.

In 794 AD Britain was invaded by the Vikings who brought with them a small Spitz-type breed of dog. Over the centuries these two breeds were

Border Collie

crossed and localities bred their own strains of Working Colley suited to the conditions. Many of these strains specialised at jobs particular to that area. One strain known as Welsh Greys were especially good at working feral goats. There was also a strain called the Welsh Hillman which was said to have been crossbred with dogs coming from North Africa. We also know that in 1514 Polish Lowland Sheepdogs were introduced into some of the working strains in Britain. Another Working Colley strain was the Highland Collie which was left unchanged until Queen Victoria in the 1800s took an interest in the breed and it gained a new popularity for a time which resulted in some crossbreeding. The Queen, always interested in dogs was also reported to have attended sheepdog trials in the late 1800s. It should however be realised that during the long period from Roman invasion to the 1800s people in Britain did not travel much and dogs in various districts remained mostly unchanged for centuries.

When nation-wide trade began to have an affect on the British farmer it also had an affect on their sheepdogs. Many of these dogs were cross-mated to produce better dogs. This also meant that over the years some of the Working Colley strains were disappearing in favour of the more efficient breeds. The Working Colley in Britain still had many different strains and these with various cross-matings produced the Bearded Collie, the Smithfield, Shetland Sheepdog, Australian Kelpie, American McNab, Old English Sheepdog, Rough and Smooth Collies (Lassie type), English Shepherd, and very likely played a strong part in some other breeds such as the Australian Shepherd, the Corgi and the German Coolie (or German Colley as it is sometimes called).

It is interesting to note that the Australian Kelpie also had its foundation with the British Working Colley with imports and crossings of a number of different strains. At a later stage the Rutherford strain was then introduced and produced the well known black dog called the Barb. The Rutherford strain was then already over 300 years old. These dogs were imported by some of the Rutherford family that migrated from Britain to Yarrawonga in Victoria and also by G.S. Kempe in South Australia.

The British Working Colley also produced one of the most noted sheepdogs in the world today, namely the Border Collie. As the name suggests the Border Collie was found around the area of the border of Scotland and England. All the Border Collies today trace back to a single dog bred by Adam Telfler from the border area. This dog was called Old Hemp and was described as being jet black with a very small amount of white on him. His coat was long and similar to a lot of Border Collies we see today.

He began to compete in sheepdog trials at twelve months of age and in his entire life remained unbeaten. It has been reported that he didn't excite the sheep but instead held them with complete control by his intense gaze. No one had ever seen a dog work sheep so well. It was also reported that he was a good-tempered dog but with a tendency to be excitable. In fact it was noted that he sometimes trembled as he worked. As a result his services as a sire were widely sought after and soon everyone wanted a pup by Old Hemp. It is estimated that he sired more than 200 male pups in his lifetime before he died in May 1901. His importance is further enhanced by the fact that his offspring also showed many of his outstanding characteristics and the line continued holding many of these traits that we now take for granted in the modern Border Collie.

Hemp's sire was a black, tan and white dog called Roy. He was described as being a plain working but easily handled dog. Hemp's dam, Meg, however was described as being very sensitive, intelligent and showed style and eye when working. She was an all black bitch. Both were from old solid lines of Northumberland Colleys.

In Britain the first sheepdog trial was held at Bala in Wales on 9 October 1873. It was reported that most of the dogs used a lot of barking while working on the course but the dog called Sam who came in third place was said to have shown eye when working. This characteristic is what we now know to be a typical trait of the modern Border Collie. The term *eye* refers to the way a dog can persuade the sheep to move by intensely looking at them instead of rushing about. This trial was won by James Thompson with his black and tan Working Colley called Tweed. She was a small black and tan dog with a white forefoot, compactly built with an intelligent foxy head and a fair coat. There were ten competitors and up to 300 spectators. For many years this was thought to be the first sheep-dog trial in the world but we now know there were trials as early as 1866 in New Zealand and a few years later in Australia.

In 1906 the International Sheepdog Society was formed. Scotland, Wales and England then had their own National Trial which was the highest award for each country. A little later Ireland also joined. The best sheepdogs in each country then competed for the highest award in the combined countries which was called the Supreme Championship. Many of the Supreme Champions were exported to New Zealand and show just how good the early Border Collies were in New Zealand. The imported dogs included: Moss 22 — later known in New Zealand as Border Boss (1907), Sweep 21 (1910 , 1912), Don 17 (1911 , 1914), Lad 19 (1913) and Glen 698 (1926).

The first Border Collie to come to Australia was Hindhope Jed in 1901. She was first imported into New Zealand by James Lilico and had already won three trials in Scotland. At the time she was in pup to another of the 'new' Border strain called Captain. She was also the first of the Border Collies in that country. She was purchased soon after at six years of age for 25 pounds by Alec McLeod of the famous King and McLeod Kelpie Stud in Australia.

Hindhope Jed was placed fourth at the Sydney Sheepdog Trials in 1902 (behind the winner, McLeod's Kelpie bitch, Biddy) and then won the event in 1903 against a field of 32. Jed was later mated by McLeod to some pure-bred Kelpies. The next Border Collie imported into Australia also came through New Zealand a year later. This was Maudie, a daughter of the famous Old Hemp. The King and McLeod partnership later also imported Moss of Ancrum, Ness and Old Bob from Britain.

The name Border Collie was used as early as 1905 by the partnership of King and McLeod in an advertisement for their dogs in Australia but the name did not become official for the new breed until much later. The Border Collie today plays a large part in Australian sheepdog trials and on farms across the country. Even though Border Collies have been bred here very successfully some breeders still import new dogs from Britain and to some extent from New Zealand to add to their bloodlines.

*Australian
Cattle Dog*

Australian Cattle Dog

Text supplied by The Australian Cattle Dog Society of New South Wales

During the early colonisation of Australia the population was mainly confined to what is now the Sydney Metropolitan area. The land holdings of this time were relatively small and the distances involved in taking stock to market were not great. The stock contained on these properties were used to seeing men and dogs around them and so were relatively quiet and controllable. Working dogs that were brought out from other countries by the early settlers, although suffering a bit from the warmer climate, are believed to have worked these quiet cattle satisfactorily.

Eventually the settlers began spreading north of Sydney to the Hunter Valley, and south to the Illawarra District. With the discovery of the pass over the Great Dividing Range in 1813, by Blaxland, Wentworth and Lawson, vast grazing lands were opened up to the west. Here land-holdings were often hundreds and even thousands of square miles and were mostly unfenced. Cattle turned loose on these properties became uncontrollable.

The most popular dog used by the early drovers and cattle owners was a working dog breed, brought out from England and known as the Smithfield. It was a big black, square bodied, bot-tailed dog, and it had a

14

long rough coat with a white frill around the neck. The head was shaped like a wedge, with long saddle flap ears, and it had a very cumbersome gait. Like the other working dogs of that time, the Smithfield found the high temperature, rough terrain and long distances too much for it.

These early working dogs all had the trait of barking and heading while working stock. This is desirable for working sheep and even acceptable with quiet cattle, but only made the wild stock on the big cattle stations rush and run off their condition. It soon became obvious that a dog with more stamina and one that would work quietly, but more forcefully, was needed to get the wild cattle to the saleyards in Sydney.

A drover named Timmins, who regularly drove cattle from Bathurst to the saleyards in Sydney, knew a lot about the native dog, the dingo, from first-hand experience. He knew that the dingo was a barkless dog, with the useful characteristic of herding its prey, then coming from behind and biting.

Timmins tried crossing the dingo with the Smithfield, with the aim of producing a silent working dog with more stamina which would be more suited to Australian conditions. This mating is believed to have occurred about the year 1830. The progeny from this mating were red, bot-tailed dogs, which were named Timmins Biters. Unlike the Smithfield, these dogs were silent workers but proved to be too headstrong, and severe with their biting.

Although this cross breeding was used for a while, it gradually died out. Other cross breeding was tried, such as the rough collie–bull terrier cross, but all these proved to be unsuitable for working cattle.

In 1840 a land owner named Thomas Hall, who owned the Dartbrook property at Muswellbrook in the Hunter Valley of New South Wales, imported a pair of smooth haired, blue merle, highland Collies from Scotland. They were good workers, but barked and headed, both undesirable traits in a cattle dog. Hall crossed progeny from this pair with the dingo, which produced silent workers that became known as Hall's Heelers. The colours of the dogs from this cross were either red or blue merle, with most of them having pricked ears and a dingo-shaped head with brown eyes, and were generally of the dingo type. Hall's dogs were a big improvement on any other available working dogs, and became much sought after by cattle men.

George Elliot, who owned a property in Queensland, was also experimenting with the dingo–blue merle Collie crosses. Elliot's dogs produced some excellent workers. A butcher named Alex Davis, took a pair of these dingo-blue merle Collie crosses to the cattle saleyards at Canterbury in Sydney. Cattle men were impressed with the working ability of these dogs, and purchased pups from them as they became available.

15

Two brothers, Jack and Harry Bagust, purchased some of Elliot's dogs and set about improving on them. They crossed them with an imported Dalmation dog in order to gain the Dalmation's sociability with horses and protection of owner's property. This cross changed the merle to a red and blue speckle. Pups were born white, developing their colouring at about three weeks of age. The only problem was that some of their working ability was lost.

The Bagusts then experimented in crossing them with the black and tan Kelpie to restore working ability. The result was a compact dog, identical in type and build to the dingo, only thicker set, but with peculiar markings found in no other dog in the world. The blue dogs had black patches around the eyes, with black ears and brown eyes, and a small white patch in the middle of the forehead. The body was dark blue, evenly speckled with a lighter blue, having the same tan markings on the legs, chest and head as the black and tan Kelpie. The red dogs had dark red markings instead of black, with an even red speckle all over. They became the forebears of our present day Australian Cattle Dog, and the blue dogs became known as Blue Heelers.

In 1893 Robert Kaleski took up breeding and showing Blue Heelers. Realising that there was no check on the judge giving the award in any way he fancied or his interest prompted, Kaleski then drew up a standard for the Australian Cattle Dog. This standard was finally endorsed by the Cattle and Sheep Dog Club of Australia and the original Kennel Club of New South Wales in 1903. The breed became known as the Australian Heeler, then later the Australian Cattle Dog, which is now the official name for this breed.

Australian Kelpie

When stock was first brought to Australia so were shepherds and their dogs. As there were no fences the stock had to be 'folded' or yarded every night to protect them from dingos, to prevent them from being taken by Aboriginals and to stop them straying. One of the earliest references we have to a particular strain of dog being imported refers to 'valuable Highland Collies' brought out on the ships *High Crawford* and *Nimrod* with cattle and horses for Sedginhoe (in the Scone district in New South Wales). These dogs could well have played a part in the development of the Australian Cattle Dog.

With the rapid increase in stock numbers and a shortage of shepherds came an increasing need for a mustering or gathering type worker dog to replace the shepherding dog used initially. There is little doubt that by the mid 1800s a great number of different strains and types of working

Australian Kelpie

dog has been imported. Some were not suited to the harsh Australian conditions but one strain, the Rutherford bred North Country Collies immediately proved ideal. This breed was named after the Rutherford family who had migrated to Australia from Scotland and once settled, were supplied with dogs by their relatives at home.

Robert Tully, a well known breeder in the Murray and Darling River areas was a strong supporter of the Rutherford strain Collie and all his stock were claimed to have carried this blood. Gerald S. Kempe was another important breeder who concentrated on these bloodlines, importing a pair, Glen and Bess, in 1894 and a further four in 1906.

Kempe was a very keen student of the breed and kept very comprehensive records. The following extract from one of his published articles 'The Collie Dog — a Tribute' describes Glen and Bess and gives a very good idea of the type of dog which has been found to be best suited to the Australian conditions.

The dog 'Glen' was a lineal descendant of 'Yarrawonga Clyde' (sire of 'Moss') and his grandfather and sire sold in Scotland for 140 and 160 pounds respectively; Mr Rutherford wrote to me that he considered 'Glen' to have been the best dog he ever saw, and well worth 200 pounds as dogs of his class were now bringing this sum in North Scotland [about 1900]. 'Bess' an all black, was the widest worker I have seen; setting off at a jog trot behind me, she almost disappeared in the distance before she would turn on any pace. She was out to find sheep and bring them home, and needed no guiding; sometimes stopping to scent, she would vet the countryside like a high class setter, putting on pace when required; at times dropping to a crouch to conceal herself until past some exposed position, careful not to disturb sheep grazing until she was acquainted with the outposts of the mob and, when gathered up, trotting back with them, sometimes 200 yards from the nearest sheep.

'Glen' was all his breeder claimed, and needless to say their blood is strong in my kennels today, and I have another pair now coming [he imported four] from Mr Rutherford of the same strain to further strengthen it. 'Glen' was a stoutly built, bold tempered Collie, shorthaired, black and tan, and with very thick head and jaws; in all round work he is likely to remain unsurpassed in my estimate, and I have had and seen the very best of them. He is equally clever at wearing down wild merino sheep off a hillside a mile or two away, or in forcing obstinate crossbreds over water.

Kempe classified the working Collies into two types — the hand or yard type which was suitable for trial competition, generally close work — and the hill or gathering dog which was best suited to the big areas and flocks. He stated that both types came in rough and smooth coats and in all colours.

The bitch which gave the Kelpie breed its name was 'Gleeson's Kelpie', bred from imported stock by one George Robertson, on Worrock station in Victoria. Gleeson's Kelpie was black and tan with semi-erect ears and a reddish tinge to her coat. It is claimed that J. D. 'Jack' Gleeson swapped the pup from Robertson for a horse.

Shortly after acquiring Kelpie, Gleeson left Worrock to travel north to take up a position on Bolero (North Belaira). In crossing the Murrumbidgee he met an old friend, Mark Tully, brother to Robert Tully, the breeder of the Rutherford strain Collies. Tully gave him a black dog, Moss, which had been bred at Yarrawonga station by John Rutherford. Moss was by Yarrawonga Clyde x Lassie, both bred from dogs imported from the Rutherford's kennels in north Scotland.

Kelpie was mated twice to Moss. The first litter whelped was shortly after arriving at Bolero. The second litter was born on North Yalgogrin and a pup from this litter was given to Steve Apps who was on Merringreen.

About this time, 1870, one Arthur Robertson brought out from Scotland a pair of black and tans for the firm of Elliot and Allen of Geraldra station. They were mated on the ship and the bitch whelped shortly after arrival. In the litter were two red pups, the rest being black and tans. One of the pups, Caesar, was given to John Rich who was on Narriah at the time. Caesar was mated to Gleeson's Kelpie and the most famous of all Kelpies was the result. When the litter arrived, a black and tan female pup was given to C. T. M. King, who entered her for the first Australian Sheep Dog Trial which was held in Forbes. Her performance there was so outstanding that the name 'Kelpie' was adopted as the name for the strain and eventually the breed.

For a short time the name Kelpie applied only to the progeny of King's Kelpie, although more properly it should have applied to Gleeson's Kelpie, for without her there would never have been a breed so named today. However soon all dogs of similar appearance were being described as Kelpies. In the entry of 35 dogs at the Sydney Sheep Dog Trials in 1889 only two dogs were entered as Collies.

The colours of the others competing in 1889 is of interest: 11 were black, 9 black and tan, 2 black and white, 3 red, 2 blue, 2 yellow and 1 black, tan and white. Several entries gave no indication of colour.

At first the name Kelpie was only adopted by owners in a relatively restricted area. Dogs of similar origin were being developed over a far wider area by the Rutherfords, Tully, Kempe, Gibson and others who used the name smooth coated Collies, even after infusing the progeny of Kelpie strain through the use of Clyde and Coil, two other well known Kelpies of the time. Strains developed in the Riverina, Murray and Darling River areas and in South Australia and Queensland did not become known as Kelpies until much later.

The first state in Australia to make provision for the Kelpie at a dog show was New South Wales where a class was provided for Kelpie, Barb or Tully. Before this they were described as either sheepdogs or Collies. This first show raised quite a storm in the Victorian *Australasian* newspaper of 1901.

> The report of a show held by Mr Court Rice in Sydney has attracted attention here, and enquiries are being made as to the origin of the 'Kelpies' included in the catalogue — I have, in answer to questions, described them as 'under-sized smooth coated sheepdogs with pricked ears' and comments were invited.

A month later another article appeared:

> A very similar dog in type and size I saw landed here from Scotland somewhere about 1866 or 1867. It came to the order of Henry Box,

who was then in the commercial establishment of Messrs. Bright Brothers and Co. of this city. I know that it was imported for station work but where it went to or what afterwards became of it I am not aware... If the so-called Kelpie is a pure bred we will have to look further back than the year 1870 for its history and to another land than Australia for its birth'.

A large section of the working Kelpie strains in Australia today can be traced directly back to the foundation stock. However the foundation of the Show or Bench Kelpie has not yet been positively established. The New South Wales strain appears to have originated from dogs of Kelpie appearance in use at the old Flemington and Newmarket saleyards rather than directly from the Gleeson, Rutherford, Elliot and Allen strain. They were interbred with the foundation strains following the establishment of the Sydney Sheepdog Trials but appear to have no other relationship. The progeny of the three foundation strains were very closely bred one with another with outstanding results in the early stages and without any evidence of loss of soundness often associated with such a high degree if inbreeding.

Many people believe that the black Barb was an entirely separate strain of dog but this is quite incorrect.

In 1902 another pair of working Collies were introduced per *S. S. Aberdeen*, and it was announced in the press that these came from the same district in Scotland as did the original Kelpies. Whilst they appear to have come from the same county there is no evidence to suggest that they were directly related to the dogs imported by Arthur Robertson.

All colours currently seen in the working Kelpies today were known in the breed prior to its coming to Australia. There is, as far as can be ascertained, no such strain as a Fox Collie in Scotland, nor up to 1990 is there any published account of a successful mating of a dog to a fox. Controlled experimental matings with dingos proved that such an infusing does nothing to enhance the value of the breed, and it appears certain that the dingo is in no way responsible for the foundation of the breed, nor can any of the Kelpie's outstanding qualities be attributed to such an infusion.

It is certain that the Kelpie sheepdog resulted from the blending of three strains of working Collies. Gleeson's Kelpie gave the breed its name, King's Kelpie's performance at the first Australian Sheep dog Forbes Trial, resulted in the adoption of the name for the strain, and later to cover all dogs of similar appearance. By 1900 the name was being generally applied to all smooth-coated prick-eared dogs, often regardless of colour and white markings.

2

Bench Standards for Working Dogs

2

The Australian National Kennel Council has outlined bench standards that which should be looked for in a good working dog. These standards were approved and adopted by the Australian National Kennel Council on 1 January, 1963.

Border Collie

General Appearance

The dog should be well proportioned with a smooth outline showing gracefulness and perfect balance. The body should be of sufficient substance to convey the impression that it is capable of enduring long periods of active duty as a working sheepdog. Any tendency to coarseness or weediness is undesirable.

Characteristics

The Border Collie is very intelligent, readily responsive to training and has an instinctive eagerness to work. Its keen, alert expression adds to its straightforward outlook. It is loyal and faithful by nature and is at all times kindly disposed towards stock.

- *Head*
The skull should be broad and flat between the ears, slightly narrowing to the eye, with a pronounced stop. Cheeks should be deep but not prominent. The muzzle, tapering to the nose, should be moderately short and strong. Lips should be tight and clean. The nose should be black and large with open nostrils.
- *Teeth*
The teeth should be sound, strong and evenly spaced, the lower incisors just behind but touching the upper.

- *Eyes*

The eyes should be oval shaped, of moderate size set wide apart and brown colour. The expression should be mild, yet keen, alert and intelligent.

- *Ears*

The ears should be of medium size and texture and set well apart. They should be carried semi-erect and sensitive in their use. The ear inside should be well furnished with hair.

- *Neck*

The neck should be of good length, strong and muscular, yet showing quality, slightly arched and broadening to the shoulders.

- *Forequarters*

The shoulders should be long and well angulated to the upper arm, with elbows close to and parallel with the body. The forelegs should be well boned and perfectly straight when viewed from the front. Pasterns show a slight slope when viewed from the side.

- *Hindquarters*

The hindquarters should be broad and muscular, sloping gracefully to the set-on tail in profile. The thighs should be long, broad, deep and muscular with well turned stifles and strong hocks, well let down. From hock to ground the hind legs should be well boned and perpendicular.

- *Feet*

Oval in shape, pads deep, strong and sound, toes moderately arched and close together. Nails short and strong.

- *Body*

The body should be moderately long, the well sprung ribs tapering to a fairly deep and moderately broad chest. The loins should be broad, deep, muscular and only slightly arched, with deep flanks.

- *Tail*

The tail should be moderately long, set on low, well furnished with hair and with an upward swirl towards the end, completing the graceful contour and balance of the dog. The tail may be raised in excitement, but not carried over the back.

- *Coat*

Double coated, the moderately long, dense, medium textured topcoat, and short, soft and dense undercoat, should afford weather resistant protection with coat on mane, breeching and brush should be abundant. On face, ear tips, forelegs (except for feather) and hind legs from hock to ground, the hair should be short and smooth.

- *Colour*

Black and white, or black, white and tan, with the black body colour being retained.

- *Size*

For dogs the desirable height at the withers should be approximately 50.8–53.3 cm (20–21 inches). Bitches slightly less.

- *Movement*

The movement should be free, smooth and tireless, with a minimum lift of the feet, conveying the impression of ability to move with great stealth. The action, viewed from the front, should be straightforward, and true, without weakness at shoulders, elbows or pasterns. Viewed from behind the quarters thrust with strength and flexibility, with hocks not too close nor too far apart. Any tendency to stiltiness, cow or bow hocks is a serious fault.

Australian Cattle Dog

General Appearance

The general appearance should be that of a sturdy, compact, symmetrically built working dog. With the ability and willingness to carry out any task however arduous, its combination of substance, power, balance and hard muscular condition should convey the impression of great agility, strength and endurance. Any tendency to grossness or weediness is a serious fault.

Characteristics

The dog's role is to assist in the control of cattle, in both open and confined areas. The ideal dog should be ever alert, extremely intelligent, watchful, courageous and trustworthy, with an implicit devotion to duty. Its loyalty and protective instincts make it a self-appointed guardian to the stockman, the herd and the property. Whilst suspicious of strangers the dog must be amenable to handling in the show ring.

- *Head*

The head should be in proportion to the body and in keeping with its general conformation. The skull should be broad, slightly curved between the ears and flattening to a slight but definite stop. The cheeks should be muscular, but not coarse nor prominent. The under jaw should be strong, deep and well developed. The foreface should be broad and well filled in under the eye, tapering gradually to a medium length, deep and powerful muzzle. The lips should be tight and clean. The nose is always black, irrespective of the colour of the dog.

- *Teeth*

The teeth should be sound, strong and regularly spaced, gripping with a scissor-like action, the lower incisors close behind and just touching the upper. They should not be undershot nor overshot.

- *Eyes*

The eyes should be oval shaped and of medium size, neither prominent nor sunken, and must express alertness and intelligence. Eye colour is dark brown.

- *Ears*

The ears should be of moderate size, preferably smaller than larger, broad at the base, muscular, pricked and moderately pointed. Spoon or bat ears are to be avoided. They should be set wide apart on the skull, inclined outwards, sensitive in their use, and firmly erect when the dog is alert. The inside ear should be fairly well furnished with hair.

- *Neck*

The neck should be of exceptional strength, muscular, and of medium length, broadening to blend into the body and free from throatiness.

- *Forequarters*

The shoulders should be broad of blade, sloping, muscular and well angulated to the upper arm, and at the point of the withers should not be too closely set. The foreleg bone should be strong and round, extending to the feet without weakness of the pasterns. The forelegs should be perfectly straight when viewed from the front, but the pasterns should show a slight angle with the forearm when seen from the side.

- *Hindquarters*

The hindquarters should be broad, strong and muscular. The rump should be rather long and sloping, with long thighs, broad and well developed, with a moderate turn of stifle. The hocks should be strong, and well let down. When viewed from behind, the hind legs, from the hocks to the feet, should be straight and placed neither close nor too wide apart.

- *Feet*

The feet should be round and the toes short, strong, well arched and held close together. The pads should be hard and deep, and the nails must be short and strong.

- *Body*

The length of the body from the point of the breastbone in a straight line to the buttocks, should be greater than the height at the withers, as ten is to nine. The topline should be level, the back strong, with ribs well sprung and ribbed back. A barrelled ribbed dog is not desirable. The chest should be deep; muscular, and moderately broad, the loins should be broad, deep and muscular, with deep flanks strongly coupled between the fore and hindquarters.

- *Tail*

The set on the tail should be low, following the contours of the sloping rump and at rest should hang in a slight curve of a length to reach approximately to the hock. During movement or excitement it may be raised, but under no circumstances should any part of the tail be carried past a vertical line drawn through the root.

- *Coat*

The weather resistant outer coat should be moderately short and straight and of medium texture. The undercoat should be short and dense. Behind the quarters the coat should be longer, forming a mild breeching. The tail should be furnished sufficiently to form a good brush. The head, forelegs, and hindlegs from hock to ground, should be coated with short hair.

- *Colour*

Blue The colour should be blue or mottled blue with or without other markings. The permissible markings are black, blue or tan markings on the head, evenly distributed for preference. The forelegs should be tan midway up the legs and extending up the front of the stifles and broadening out to the outside of the hindlegs from hock to toes. Tan undercoat is permissible on the body providing it does not show through the blue outer coat. Black markings on the body are not desirable.

Red Speckle The colour should be of a good even red speckle all over, including the undercoat (not white or cream) with or without darker red markings on the head. Even head markings are desirable. Red markings on the body are permissible but not desirable.

- *Size*

The desirable height at the withers should be within the following dimensions : dogs 45.7–50.8 cm (18–20 inches), bitches 43.1–48.2 cm (17–19 inches). Dogs or bitches over or under these specified sizes are undesirable.

- *Movement*

Soundness is of paramount importance. The action must be true, free, supple and tireless, the movement of the shoulders and forelegs with the powerful thrust of the hind quarters should be in unison. It is essential that the dog be capable of quick and sudden movement. Stiltiness, loaded or slack shoulders, straight shoulder placement, weakness at elbows, pasterns or feet, straight stifles, cow or bow hocks, must be regarded as serious faults.

Australian Kelpie

General Appearance

The general appearance should be that of a lithe, active dog of great quality, showing hard muscular condition combined with great suppleness of limb, and conveying the impression of being capable of untiring work. Any indication of weediness is a serious fault.

Characteristics

The Kelpie is extremely alert, eager and highly intelligent, with a mild, tractable disposition, inexhaustible energy, loyalty and devotion to duty. It has a natural instinct and aptitude in the working of sheep, both in the yard and open country.

- *Head*

The head should be in proportion to the size of the dog. The skull should be slightly rounded and broad between the ears, the forehead curving very slightly towards a pronounced stop. The cheeks should be neither coarse nor prominent, but round to the foreface, which is cleanly chiselled and defined. The muzzle should be of moderate length, tapering towards the nose and refined in comparison to the skull. Lips should be tight and clean. The nose colouring should conform to that of the body coat.

- *Teeth*

The teeth should be sound, strong and evenly spaced, the lower incisors just behind but touching the upper.

- *Eyes*

The eyes should be almond shaped, of medium size, clearly defined at the corners, and show an intelligent and eager expression. The colour of the eye should be brown, harmonising with the colour of the coat. In the case of blue dogs a lighter coloured eye is permissible.

- *Ears*

The ears should be pricked and of moderate size, running to a fine point at the tip. The leather should be fine but strong at the base, inclining outwards and slightly curved on the outer edge. The inside of the ears should be well furnished with hair.

- *Neck*

The neck should be of fair length, strong, slightly arched and gradually moulding into the shoulders, free from throatiness.

- *Forequarters*

The shoulders should be clean, muscular and well sloping with close set withers, the upper arm slightly angulated to the forearm, the elbows set

27

parallel to the body. The forelegs should be muscular with strong but refined bone, perfectly straight when viewed from the front, but pasterns should show only a slight angulation to the forearm when viewed from the side.

- *Hindquarters*

The hindquarters should show breadth and strength. The rump should be rather long and sloping, and laid at a corresponding angle to the shoulders, the stifles well turned, the hocks fairly well let down, and placed parallel with the body.

- *Feet*

The feet should be round, strong, deep in the pads, with close knit well arched toes and strong short nails.

- *Body*

The chest should be deep rather than wide, ribs should be well sprung (not barrel ribbed) with firm level topline, strong and well muscled loins and good depth of flank. The length of the dog from the point of the breast bone in a straight line to the buttocks, should be greater than the height at the withers as ten is to nine.

- *Tail*

The tail during rest should hang in a slight curve. During movement or excitement it may be raised, but under no circumstances should the tail be carried above a vertical line drawn through the root. In length the tail should reach approximately to the hock.

- *Coat*

The outer coat is moderately short, flat, straight and weather resisting, with a short dense undercoat. On the head, ears, feet and legs the hair is short. The coat is longer at the neck, showing a fair amount of ruff, and at the rear of the thighs, forming a mild breeching. The hair on the tail should be sufficient to form a good brush.

- *Colour*

Colours are black, black and tan, red, red and tan, fawn chocolate, and smoke blue.

- *Size*

The height at the wither should be 45.7–50.8 cm (18–20 inches) for dogs, 43.1–48.2 cm (17–19 inches) for bitches.

- *Movement*

It is essential that the Kelpie be perfectly sound, both in construction and movement. The gait is four square, smooth, free and tireless, with the ability to turn suddenly at speed, and as well be capable of the crouching stealthy movement demanded by its work. Any tendency to cow hocks, bow hocks, stiltiness, loose shoulders or restricted movement, weaving or plaiting are serious faults.

28

3

Selecting a Pup

Great care should be taken when selecting a pup. It is most important to select one to suit your own personality.

In general the bold, friendly pup is the better type of dog to select. If you are outgoing and tend to use a loud voice this bolder pup would probably be more suitable to you. However, if you're a softer spoken person who handles dogs quietly, then a softer or quieter dog would suit you better.

Selecting a pup less than three months old is more difficult as the pup will not have yet grown accustomed to people or the work you intend it

Pups should be selected from well bred dogs

to do. In general, the best method of selecting a really young pup is to choose one which is bold and friendly, and which is wide between the ears and eyes. This can be called a 'good type of pup'.

Take time to select your pup. I'm sure that when faced with the selection of a pup from a litter you'll instinctively be able to judge the most appropriate dog for you.

A pup should always be bought from a well known breeder of working strain dogs, such as breeders who are members of the Working Kelpie Council of New South Wales or the Sheepdog Workers of New South Wales.

There are several good reasons for this:

- The chances of getting a first class dog with top working ability is greatly increased because the better breeders are very careful in the selection of their breeding stock. Their percentage of failure is very small. Their reputation depends on it.
- Most good breeders will give a guarantee to replace the pup if its work is unsatisfactory.
- Your chances of getting a bargain by buying a cheap pup out of the neighbour's old bitch are not very great. What usually happens is the pup is cared for and fed for 12 to 18 months before you find out that at best it is second rate, at worst no good at all.

I would recommend that when selecting a pup you keep uppermost in mind what will be required of it. Consider whether the dog will be used for yard, paddock, cattle work, sheep work or as an all rounder with sheep and cattle.

Then go to a breeder who specialises in breeding the type of dog you need. Remember that some breeders have different lines of dogs that are bred for various specific tasks. Most good breeders will be able to advise and help you.

The pup selected should be of good type for its particular breed. A bold friendly pup is usually preferred to a timid shy one. Look for conformation faults like overshot or undershot jaws, light bones, turned or twisted feet and legs. None of these faults are desirable and good stud breeders shouldn't offer them for sale. Feet and legs are of great importance in all working dogs, but especially so in long casting mustering dogs that will be often working in extreme heat and harsh conditions.

It is important to choose a pup with an agreeable temperament

4

Initial Training

The process of training dogs, although rewarding, can be a long and sometimes frustrating task, but the end result is usually pleasing and worthwhile. Bringing a pup up from being a playful nuisance, seemingly always underfoot or in the wrong place and watching it develop gives much pleasure and enjoyment but also much frustration.

The purchase of a pup is the beginning of this long and enjoyable working relationship. The selection, bonding and initial training is a most important phase to ensure success.

Bonding

As much human contact as possible is essential with young pups. So after selecting your pup it is important to keep it with you as often as practicable. This helps form a lasting bond between the pup and you. The bond formed during this stage stays with the dog all its life. If you are lucky enough to have children they can help immensely with the bonding process. If the children play with the pup, handle it and treat it kindly and gently an excellent job will be done to familiarise it with humans and to cement a lasting bond with its master. A newly purchased pup which is left tied up with the other dogs will only learn to bark at the end of a chain.

Discipline

The old adage 'a smack when they're naughty and a pat on the head when they're good' applies pretty well to dogs. If a dog is treated in this manner, handlers will find they will not have a lot of trouble.

Basic obedience training is the key to success. Remember, dogs are pack animals and you need to take the place of the pack leader. If the dog is taught to obey and understands this from an early age it is highly

Keeping your pups with you helps bond them

unlikely any heavy handed discipline will be required. However sometimes instances will occur when discipline will need to be applied. An important consideration when disciplining a dog is timing — the dog needs to know what it has done wrong. A common mistake is to command the dog to **come** and then hit it for something it had done wrong earlier. The dog thinks it is being punished for coming to the handler.

When a situation arises when the dog needs to be punished the use of a stern voice may be all that is required. This should always be your first option. Dogs will respond to kindness and this should have been reinforced during the bonding process. They want to please their master and

35

a stern voice should do the trick as long as they understand what they have done wrong.

If heavier handed discipline is warranted lifting the dog off the ground by the scruff of the neck, giving it a shake and verbally chastising it usually proves the most effective method.

If the dog proves to be incorrigible and you believe it needs to be hit *do not under any circumstance use a heavy object that could cause permanent damage to the dog*. Remember the purpose of discipline is to instil in the dog that you are the pack leader. Continual heavy handed discipline does no more than undo the bonding you have worked so hard to create. So if you as a handler find you are hitting the dog on a regular basis go back to the beginning, run through basic obedience again and correct the problem.

If your dog appears to be totally disobedient, instead of blaming it take time to look closely at your own methods. The problems could lie with you and not the dog. It may not understand what you require of it. Patience is the key, basic obedience is the tool, a thorough grounding in both is essential.

Early Training

Pups under six months of age have a very short concentration span so the training should be kept simple. An easy and important lesson is to teach the dog to come to you.

Pups learn through association and reward. If you call the pup by its name as it is being fed and pat and praise it when it comes to you it will associate its name with food and kindness.

Rewarding dogs with food is not recommended as they grow older. Rewarding dogs with food at any time after the age of 15 weeks leads to very bad habits for the dog can and will expect food as a reward every time it does something for its handler. Working dogs, especially Kelpies, Cattle Dogs and Border Collies, are just that, working dogs, they work because they enjoy it. The only reward a working dog needs is to know that it is pleasing its master or mistress so when it performs the way you want it to, make sure you let your dog know you are pleased with it.

Commands

Before we start with basic obedience training it is important to discuss the way commands are given to the dog for the commands will stay with

with it throughout its entire working life. At all times commands should be given in a quiet and calm manner and *a distinct difference should be detected between a command and a reprimand*. A good point to remember is that dogs understand tones but do not understand language, thus it is most important to be totally consistent in tone and inflection when giving commands for each action. Any variation in command will only confuse them.

Remember *talk to your dog, don't shout at it* or you will always have to gain it's attention by shouting.

The voice command can be substituted for or combined with a whistle or a hand signal or both. Incorporating all three when training will prove very useful later when the dog is working at a distance from the handler.

To include all three in your training regime is relatively easy. For example, when giving the **sit** command show the dog a raised open hand and after giving this voice command give a short sharp whistle. The dog will associate all three — the hand, the voice and the whistle — with the command **sit**

Whistle Commands

A whistle command is extremely useful when controlling your dog from a distance and can be used in conjunction with hand and voice commands. Whistle sounds can be made from a variety of sources and all seem to work as well as each other. I have seen one of the better three sheep trial men in Australia use a football referee's whistle with great success.

Some dogs respond to whistles better than either hand or voice commands. Some of the top trial handlers use multiple whistle tone and combinations to control their dogs.

However, for general farm and station work I have found three or four different whistle commands adequate. One whistle for a recast at a distance, another different sound to sit or stop a dog, another to call the dog to you. The style or tone or number of whistles given is unimportant provided each is different and distinctive. It is very important to use the same whistle for the same command all the time to avoid confusing your dog.

More than one command for the same action is commonly used by professional trainers preparing dogs for sale. As an exporter of dogs to the United States I train dogs to respond to voice, hand and whistle signals because I have found the dogs have trouble determining a command given with differing American voice tones. By training dogs to understand hand and whistle commands this problem is easily overcome.

Teaching a Dog to Bark on Command

A dog that will bark on command is a very valuable asset to any stock handler. If it learns to respond to commands to go left and right and to look for stragglers it is even more of a gem.

There are two methods of teaching a dog to bark on command. The method I strongly recommend is a positive method using excitement, the other is a negative method, instilling fear.

Through observation it will become apparent to you that the dog becomes excited when certain things happen. It may bark at the sound of an excited voice, the rattling of a tin can or a chain. When it is barking give the **speak up** command and repeat this command every time the dog barks. After a period of time it will learn to bark on command. When the dog actually barks on command reward it to reinforce the response you have received.

The second method, and one I certainly don't recommend is to get the dog so frightened of an object (usually a whip) that the dog barks out of fear. This method is exceptionally cruel and serves no positive purpose and is best left alone.

Some dogs (the strong silent types) will never be encouraged to bark no matter the method used. If this is the case with your dog accept the fact it will still make a good yard dog and save your ears.

Chaining the Dog

Chaining your dog is an essential part of its overall training. A working dog must behave while it is on the chain. A pup must be prepared for this gradually. The first step is to place a small tight collar on the pup and once it is comfortable wearing this the second step is to fasten a short piece of light rope or a lead to the collar. The pup will then become used to dragging the lead around.

These simple steps prove themselves invaluable when the pup approaches 15–16 weeks and is now ready to spend short periods of time on the chain. I stress *short periods* of no more than 10–15 minutes duration once or twice a day at first, then for gradually longer periods as the pup becomes more used to being restrained. Under no circumstances leave the pup tied for long periods of time.

Your pup needs to be broken into a chain at 15–16 weeks old. This starts the process of gaining control over the excitable pup. If a dog isn't chained until it is 10–12 months old it can develop into an almost uncontrollable and usually disobedient animal. Another point to remember is

*What did he leave
me here for?*

that most of your pup's initial training requires the use of a lead so if the
pup is used to being tethered it will make this learning stage a little easier.

From age 3–5 months the pup must be taken with you as often as possible. While the pup is still young (3–4 months) put it in the cab with you —
this also nurtures the bonding process. When the pup reaches approximately
4 months take time to introduce it to the tray of the truck. Remember
special care should be taken to ensure that it doesn't come to grief. I have
received many phone calls telling me the pup has been killed because it
has fallen, jumped, or choked itself by falling from the tray of the truck.

All young pups must be chained to the truck and chain lengths should
be no longer than 30 cm, securely attached directly behind the cab. Three

39

chains are more than enough. This instils confidence in the pup and teaches it to maintain its balance in a totally alien environment. Care should be taken when chaining your pup to the backs of trucks which have steel or aluminium trays, as they get very hot in the summer sun and can burn the pup's feet.

First Introduction to Stock

Young pups of 15–16 weeks old are ready for their initial familiarisation with stock.

Pen a few quiet sheep in a large yard, let the pup go and watch closely what happens. I stress 'watch closely' because if the sheep trample the pup to death the training program is over. The pup's actions when confronted with livestock for the first time give no real indication as to its potential, so don't be disheartened if it does nothing at all, or if it chases the sheep madly round the yard barking its head off. Most young pups of any breed, even non-working strain dogs, will chase anything that moves. If your pup quietly watches the stock and attempts to cover when a sheep breaks this indicates some potential but it is not really significant. Three-month-old pups are likely to do anything, so don't worry if the pup doesn't work at all. If your pup does nothing or if it works well don't attempt to control it. At this very early stage let your pup enjoy itself. Give it time to gain confidence, for it is a very important stage of its development into a working dog.

5

Basic Obedience

(Come, Stop, Sit, Down, Stay)

5

Basic obedience training should begin when the pup is approximately five months old. Most experienced handlers prefer this age, some start earlier, and some very good handlers don't teach basic obedience until the dog is actually working. I would not recommend the latter method to novices, as gaining control of an older dog is much more difficult. A young dog at five months has a much improved concentration span and when it is introduced to stock and starts serious training, it will be easier to control.

Basic obedience should be taught well away from anything and anyone that could distract the young dog, that is away from children and other dogs. Lessons should be kept short , 10 -15 minutes maximum, or the young dog will lose concentration.

Come

The **come** command is fundamental in a young dog's training. However many dogs aren't taught this basic instruction. The benefits of having a

A light tug on the lead is sometimes necessary

dog that comes on command are too numerous to mention. If your dog was trained to **come** at feed time when a young pup it will already be used to responding to this command. However if it refuses a different method can be applied.

Tie a 5-metre length of rope to the dog's collar, call it and give the rope a sharp tug. When it comes to you reward the pup generously with a pat. Eye contact with the pup is also important and it may help the training by squatting to be at eye level position with the pup. If the pup still refuses to obey a firmer tug on the rope may be necessary, followed by lots of praise. Always treat your dog kindly and it will respond.

Stop

The **stop** command is probably second only to the **come** command in terms of usefulness. If your young dog will come to you when called and will stop when commanded, the handler will find the young dog much easier to train because the handler is in control.

Teaching your dog to stop is reasonably easy and the young dog normally responds quickly. When your dog is consistently coming to you on command, position yourself 10–15 metres away from the dog and give the **come** command. As your dog is running toward you raise your hand

When the lead tightens your dog soon learns to stop

and give the command **stop**. If your dog stops praise it, then move back a few paces and call it to you again. At first the dog will show no response, but perseverance usually gets results.

Another method utilises the short rope, and a young pup's spirit of adventure. Attach the rope to the dog's collar and when the dog starts wandering away from you give the **stop** command. If your dog stops praise it, if it doesn't, take up the slack in the rope give the **stop** command again and tighten the rope. Your dog will soon learn to stop.

Sit

The **sit** command is used frequently with an advanced working dog. During training this command is used to gain control of the pup as it is working. Many methods may be employed to train a pup to sit.

A common method is to secure the dog with a rope or a lead and keeping the rope semi-tight push the dog's rump to the ground while giving the **sit** command. If the dog sits reward it with a pat. Repeat this process until the dog is sitting on every command.

Light pressure on the rump does the trick

Down

The **down** command is usually employed to gain tighter control of the dog and the mob it is working.

   To train the pup to lie down, stand close to the pup with the rope or lead in your hand, then step on the lead so it can run under your instep. Now give the **down** command and pull on the rope while pushing the pup to the ground. Reward the pup with a pat when it lies down. Border Collies usually perform this action instinctively while most Kelpies prefer to stand whilst working. Either working style is acceptable depending on the handler's preference.

Sit at a Distance

The next step is to teach your dog to sit on command at distance. It is extremely desirable to have your dog sit when commanded to do so even if it is 50 metres away from the handler. A well trained dog will sit at distances in excess of a kilometre from the handler.

Training the pup to lie down

Start initial training by sitting the dog 5 metres from you, and once it has learned to obey gradually increase the distance between you and your dog. If you encounter problems with the dog wanting to return to you before sitting, it will be beneficial to train the dog to **stay** first.

Training the Pup to Stay

The **stay** command is introduced to the pup when it is continually responding to the **sit** command. It is invaluable to have a dog that will stay on command, especially when working weak or pregnant stock.

The **stay** command is taught by facing your pup and giving the **sit** command. Then slowly back away from the dog with your hand raised while giving the **stay** command. Once again perseverance is the key to success. Usually when you begin using the **stay** command one of three things happen. The first but rarest could be instant success with the dog sitting and staying and not moving until told.

It is more likely that the dog will stand up and follow you or stand up and run off in the opposite direction. If the pup follows you immediately give the **sit** command and the pup should sit. Once the pup is sitting,

Control allows a handler to position himself correctly

Training your dog to stay

back away from it again giving the **stay** command. You may have to repeat this process until the pup learns.

If the pup runs off in the opposite direction tie a piece of rope or a lead to its collar. A light tug on the rope is given when the pup starts to run off, or alternatively stand on the rope while giving the **stay** command.

If a pup can be commanded to stay at a distance of 20-40 metres from a mob, the stock can move at their own pace, placing them under less stress.

Training the Dog to Jump

When your dog is performing its basic obedience consistently a new command can be included in its training. The **up** command is given when you want your dog to jump up onto an object.

It is useful to introduce this command at this stage of training because the dog will need to know the command when it is learning to back sheep during yard work.

The dog will also be required to jump onto the back of the truck on command.

Although the dog won't be required to perform these tasks at this stage of training, it is the ideal time to introduce the command to the dog.

The most successful method I have found to teach a dog to jump, is to use a lead or a piece of rope and attach it to the dog's collar. Find a raised flat surface approximately 60 cm high and while holding the lead climb onto the platform and call the dog with the **up** command. Reward the dog when it jumps up, repeat the steps above until the dog consistently performs the task.

Now change your position and stand on the ground next to the platform, give the **up** command again and reward the dog when it performs the task. When the dog consistently obeys, the height of the platform may be raised.

A useful method of instruction

The eventual result of correct training

A mature and well trained dog can handle a large mob

6

Working Sheep

6

Once your dog understands and obeys the basic commands of **sit, stay,** and **come** and has been initially familiarised with stock, more intense work training can be undertaken. Debate among experienced trainers as to the best age to introduce dogs to stock work has gone on and probably will go on for ever. Some believe that as soon as a pup shows interest in work it should start. Others believe dogs have to mature to 8 to 10 months and in some cases even older before they are mentally and physically strong enough to learn and control stock.

To illustrate this point more clearly take the case of a keen immature

Some pups will work at an early age

pup and a small mob of sheep. Because the pup can't run fast enough to head the mob and block them, it only manages to get to the wing of the mob. The mob starts running away and the pup chases them down the paddock barking its head off as it goes. Not only does this encourage bad habits in the pup, it can undo much of the basic training given in previous months.

I would recommend starting this phase of training with a young dog of 5 months. Select a small mob of 15–20 quiet sheep that are used to being worked by dogs. If no sheep are available a small mob of young cattle can be used; however I strongly recommend the use of sheep at this early stage.

Place the sheep in a small paddock. (Too large a paddock will allow the sheep to get too far away.) Take the young dog into the paddock and walk toward the sheep and when the sheep start to move, the dog (if it is ready to work) will want to run toward the mob and should run to the opposite side of the mob to the handler. As the dog is running toward the mob give the command you will use to cast it — **go back** or some other appropriate word.

In all probability the young dog will ring the mob, that is he will run right round them and back to the handler. This happens instinctively with a heading dog as they will constantly want to be on the head of the mob.

Ringing should be discouraged at this stage by giving the **sit** command when the dog is on the opposite side of the mob to the handler. If the young dogs basic obedience training has been thorough and effective it should sit and the mob will move away from it and toward the handler. With the mob moving the dog will once again want to move to the front or head of the mob. If this occurs and it probably will, stop the dog at the three o'clock position on the mob. Once stopped the dog will most probably want to run behind the sheep and attempt to head them on the other side. Stop the dog at the nine o'clock position on the mob. With the action of stopping the dog on the wings of the mob and encouraging it to **go back** behind the mob the dog is in fact moving the mob toward the handler. Start moving backwards constantly watching the dog and encouraging it to bring the mob toward you.

It is important to keep this lesson short (5–10 minutes). Call your dog off the sheep with the **come** command and reinforce the lesson with lots of praise.

You have just witnessed the perfect training session for a small cast, lift and draw. In fact it won't be that easy. Most probably as soon as the dog sees the sheep it will charge at the mob like a bull at a gate and continually ring the mob, totally ignoring every command screamed at it. At this stage I welcome you to the joys of dog training!

When this happens stay calm and get as close as possible to the dog. You may have to block it at the wing of the mob and very firmly issue the **sit** command. Provided you are close enough to the dog it will probably obey. Having gained control of the dog repeat the procedure but be more firm this time.

Discipline should be limited to verbal scolding at this stage. *Under no circumstances hit the dog*. Physical intimidation could actually stop the dog from working.

On the other hand the young dog when released may look at the mob, look back at the handler and then wander off sniffing the daisies, totally disinterested in anything that is going on. If this is the case the handler has to attempt to interest the young dog in the task ahead. The handler can try some of these suggested methods.

- Call the dog to you and walk with the dog toward the mob. As the mob starts to move away this movement may encourage the dog to run to the head of the mob.
- Tie the young dog up in the yard and let another dog work. This method is most effective while yard work is being performed.
- Working the young dog with an older dog could encourage it to start. Don't do this too often though as the young dog could become dependant on the old dog and will not work unless the old dog is there. This is a last resort training method.

This stage of training can be extremely frustrating, however keep in mind many of the very best dogs didn't start to work probably until they were 12–16 months old. If your dog won't work you must remember, training is a repetitive process that takes time. Continue bonding the dog with you, continue its training and every now and then reintroduce it to stock.

If after a fair period (17–19 months) you have no success contact the breeder. Most reputable breeders will replace the dog.

The Working Kelpie Council of Australia Inc. issue the following recommendations to their breeders.

To guarantee to replace, or refund the purchase price of a pup which fails to develop a desire to work by the age of 12 months or which exhibits an inherited physical condition which effects its ability to work subject to the following conditions:

The dog or pup is returned in good order and condition at the buyer's expense or other mutually agreed to conditions are fulfilled.

- In the case of demise, or necessary euthanasia by a qualified person, that the cause of the loss can be associated with exposure to or the existence of a condition prior to the time of dispatch. In this circumstance the breeder reserves the right to request the claim be accompanied by a certificate signed by a registered veterinarian or other suitably qualified person.

- It can be established that whilst under guarantee the pup/dog had been properly housed, handled, fed and routinely treated for internal and external parasites and protected by routine vaccination against commonly associated diseases such as canine distemper.
- It can be established that the pup/dog was given adequate exercise, training and opportunity to develop its inherited working instinct or in the case of started or broken-in dogs they were given adequate time and opportunity to adjust to their new environment.

While carrying out this early training with stock, several points must be kept in mind.

- *Never over control your dog*, remember it is just learning. Applying too much control can confuse a young dog to a stage where it will not work at all.
- *Never expect too much of it.* Allow the young dog to develop at its own pace.
- *Don't attempt to teach it to drive.* Driving sheep away from the handler is not instinctive. Forcing the dog to drive too young may inhibit its ability to head the mob.
- *Never let the dog cut sheep from the mob.* Cutting sheep from a mob is a very bad fault and should be stopped immediately. Command the dog to sit and severely verbally chastise it. (Discipline is dealt with in more detail in Chapter 4.)
- *Stay calm.* When things go wrong, stop the dog, calm it down, calm yourself down and start again.
- *Finish on a positive note.* Always finish a training session on a positive note. When the young dog has done something right, sit it, praise it and lead it away from the mob. Take it back to the kennel and chain it.

Casting

The cast describes a dog's ability to run in a wide arc or curve around the stock.

Casting is the development of a dog's instinctive heading ability and is used primarily for gathering or blocking stock. Teaching a dog to cast is merely controlled development of a natural ability.

At present the only casting your young dog has done will probably

Casting to the head

have been very close (20–30 metres) and instinctive. The handler has had no control of speed or distance. Speed is most important as the dog should get behind the mob as quickly as possible. This doesn't mean, however, that the dog should go flat out and get itself so excited that it chases the mob away before it achieves its correct point of balance directly opposite the handler. The correct distance or width from the mob your dog should cast depends on the environment it is working. When casting your dog in hilly country where the mob are spread out, the ability to cast very wide and virtually run the fence line is essential to avoid missing any of the mob. If however, you are grazing heavily timbered country it is an advantage if the dog casts relatively close to the mob.

It is worth mentioning that breeding determines to a large degree the dog's aptitude to be a top class musterer. The wide sweeping cast and the dog's ability to re-cast (cast itself out again if it sees another mob further away) is very difficult to teach a close casting dog. The great ability of wide casters is mostly genetic.

All is not lost however, casting faults can be corrected. A couple of techniques I have used with good success require some perseverance and patience.

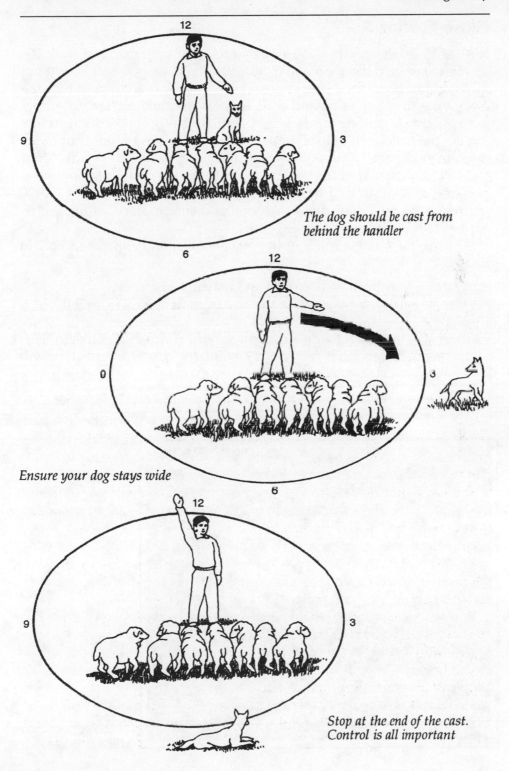

The dog should be cast from behind the handler

Ensure your dog stays wide

Stop at the end of the cast. Control is all important

Close Casting

Muster 15–20 sheep into a 2–3 acre paddock. Sit your dog and walk 10–15 metres toward the mob and as close to the mob as possible without moving the sheep. Face your dog and give the **away** command for anticlockwise direction, point and walk in the opposite direction encouraging your dog to cast. As the dog casts it will probably attempt to run close to you, the handler. If it does chase it out and give the **get out** command.

Remember not to continually cast your dog in one direction. Alternate the direction or your dog will become used to casting in one direction and may never cast in the other. Keep the initial cast short (20–30 metres) and when your dog achieves that distance, start casting a little further (40–50 metres).

During the lessons some of the responses listed below are likely to happen.

- The dog may become confused and may not cast at all.
- The dog may cast but when the handler attempts to chase it out it stops.
- Your dog may be a hard or sticky eyed dog (has too much eye). This means the dog's stare becomes so intent, the dog won't hear, or will ignore, your commands.

If your dog refuses to cast it needs reassurance and the handler may

Sitting on completion — success with initial training

need to stand closer to it. You can extend the distance once the dog realises why you are standing way down the paddock.

If your dog stops, it does so because it believes it has done something wrong. Reassure your dog as you chase it wider.

Positioning yourself correctly also helps a great deal when teaching your dog to cast wider. Placing yourself beside your dog and between it and the mob seems to get it to see that you want it to run wider

However, if your dog cuts into the mob instead of completing its cast, you need to position yourself further to the wing of the mob, at the 9 o'clock position.

If your dog casts wide when you are close to the mob but close casts when the mob is at a distance you will have to cast the dog while standing closer to the mob. You will also be closer to any problem that might happen.

Always use the **get out** command every time the dog is forced wide of the mob. This will allow you to give the **get out** command from a distance at a later time when the dog is performing well.

If your dog casts too wide, you can consider yourself very fortunate as these dogs make beautiful mustering dogs. Some of the great Kelpies and Border Collies were this type of dog.

A big fault with this type of dog is if after a wide cast, the dog refuses to move in on the mob to lift them (start them moving slowly). This problem is much more difficult to fix than that of a fast running close casting dog. It seems to be a genetic problem with the dog's makeup. Quite a lot of these dogs are weak (lacking force) and sometimes can actually be frightened of the sheep.

All is not lost — though there are certain methods that can be employed.

- Keep all casts short while the dog is young, this will allow it to gain confidence, while it is close to the handler.
- Work the dog in big yards, and in forcing situations to develop confidence.

Never discipline the dog for not forcing, it will only make things worse. The dog needs positive reinforcement, not hindrance.

Sticky eyed dogs when cast will often stop short (some may only cast 20–30 metres) and stare at the sheep or stop on the wings of the mob at the 9 o'clock or 3 o'clock position, or they may complete their cast but have such strong eye that they cannot initiate a lift.

This type of dog can cause considerable problems for the inexperienced handler. It is most important to break its concentration as quickly as possible. The handler can give the **get out** command and chase the dog wider of the mob or crack a stockwhip to break the dog's concentration.

When its concentration is broken recast it and keep the dog moving every time it looks like stopping give the command you use to get the dog to cast.

Lift and Draw

The *lift* refers to the first movement of the sheep and should be created by the dog moving in slowly. Some dogs bounce in with a mad rush and have sheep running in all over the place, effectively spoiling a good cast. Good control of your dog will prevent this happening, this is why a dog is taught to sit at the end of its cast. When the dog is sitting at the end of its cast, give it time to settle down, 10-15 seconds, more if necessary then give the command to **come up** and the dog should lift the sheep gently.

The *draw* is the action of bringing the mob to the handler, once again this action should be performed calmly and efficiently, neither too fast or too slow.

Initial movement

Lift should be slow and gentle

Draw must be controlled

If your dog is moving the mob too quickly use the **sit** command. This will allow the dog and the mob to calm down. If the mob is moving too slowly you will need to excite the dog with an encouraging command or a whistle.

Direction

Many stockmen work their stock quite satisfactorily without teaching their dogs to respond to **go right** or **go left** commands. Most will stop their dogs first and then send to try and direct them to go in another direction with a hand signal. In many instances this is adequate. However, a dog trained to respond to directional commands leaves the stock, the handler and itself much less agitated at the end of the day.

I have found that the most effective method of teaching direction takes advantage of the dog's natural instinct to balance the mob with respect to the handler. Most dogs working sheep will balance or hold the mob directly opposite to where the handler is standing. If you imagine the clock face with the handler standing at 6 o'clock and the mob in the middle of the of the clock face when balancing the mob the dog will be standng at 12 o'clock. Making use of this instinct the handler moves clockwise to 9 o'clock and the dog will move clockwise to 3 o'clock. To regain the balance of the mob the dog has moved in a clockwise direction without being given a command. If a command such as **bye** or **over** is given as the handler is moving to a different position the dog will learn to respond to that command.

When teaching your dog to move in an anti-clockwise direction on command it is advisable to select a command that sounds totally different to the clockwise command. A couple of commands you could use are **away** or **back**. To start training the dog position yourself at 6 o'clock and move anti-clockwise to the 3 o'clock position. As the dog moves to achieve its balance give the command you have chosen for the anti-clockwise direction.

Persistence is the key. Your dog will eventually learn the command and move in the direction you require on command.

It is of great benefit to teach hand signals at the same time as you are teaching voice commands for direction. When your dog is working 2 kilometres away out of earshot a directional hand signal is invaluable. Hand signals are quite easily taught by including them in the clockwise/anticlockwise training technique. When the verbal command for clockwise is given raise your right arm until it is horizontal to the ground. The dog will soon learn to respond to the visual command as well. The anti-clock-

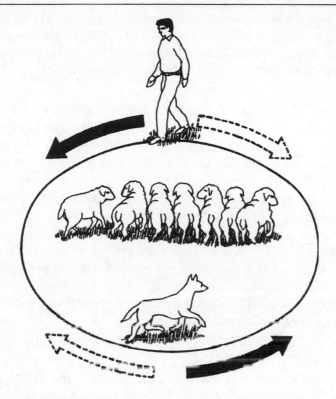

Using natural balance instinct to teach direction

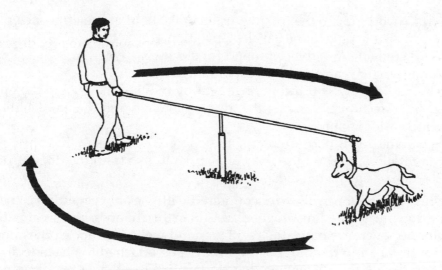

A rotating pole is an effective method of teaching direction

wise hand signal can be given with your left arm held horizontal to the ground as the verbal command for anti-clockwise is given.

There are other quite successful methods of teaching direction to dogs. These methods are taught when the dog is 6-7 months old. No sheep are needed when using these methods:

- Get a piece of dowel or a light rod approximately two metres long.
- Attach a lead 1-2 metres long to the rod and the dog's collar.
- Place a bale of hay or a 44 gallon drum between you and your dog to prevent it coming directly towards you and also give it something to run around.
- Next give the clockwise command of your choice and move the stick with the dog attached in that direction.
- Change your position and repeat the procedure.
- When the dog is responding to the command reverse the direction by giving the anti-clockwise command continue until the dog is responding to both commands.
- Remember with all training give plenty of praise and pats when your dog performs the task required.

Another very effective method utilises a constructed device known as a rotating pipe. This device consists of a length of pipe approximately 3 metres long pivoted in the centre and anchored to the ground. One end of the pipe has a half-metre chain and clip attached to it. This method is used by some trainers of trial dogs because not only will it teach direction, it will also encourage a dog to run in an arc. To use this method to train direction:

- Attach your dog to the rotating frame, take hold of the opposite end and walk the pole to the 9 o'clock position.
- While moving give the command for clockwise **bye**.
- A hand signal can also be incorporated at this time.
- The anti-clockwise direction is taught by moving the pole in the opposite direction and giving the **away** command and the hand signal you consider appropriate.
- When the dog is performing the directions well move to the centre of the rotating pipe and use the same apparatus to train the dog to run 360 degrees on command.

The major problem I have encountered with the above method is that some dogs object to being dragged around in this manner. However with a little perseverance most dogs will respond well. A rotary clothes line works as well as a rotating pole. With the dog attached to one side and the handler on the other move the clothes line to achieve the same results.

64

The ability to cast either direction is a strong advantage

Should you choose to use this method I give you a little extra advice learnt through personal experience. Ensure your wife isn't around and make sure the clothes are off the line because the dog finds it difficult to see your hand signals through the sheets and undies!

Eye

Eye in a working dog describes its ability to hold or control stock with its eyes, very much the same way gun dogs will point at hiding game. Eye is believed to have been introduced from gun dog strains during the working dog's early development. Eye with working dogs is very important, it allows a working dog to hold and balance stock, it affects to some degree a dog's heading ability and its keenness to work.

The eye in working dogs can be placed generally into three categories:

- Too much eye (hard eyed or sticky eyed).
- No eye (loose eyed). No ability to hold the sheep with their eye.
- Dogs that have the correct amount.

Hard eyed dogs have a very ineffective working ability. They tend to tie the sheep down with their eye, to the point when the sheep won't move. The very worst of these dogs concentrate their gaze on one of the mob totally, unaware of anything else going on around it, so much that the rest of the mob may wander away. Severely hard eyed dogs will not hear commands (or they will choose to ignore them) and when they finally decide to cover the rest of the mob it is usually too late. This type of dog is usually a short caster (they often stop half way to the mob or as soon as they achieve direct eye contact with the mob).

This problem is mentioned in the section on casting and can be a major problem for an inexperienced handler. The corrective methods listed below will help the handler overcome this problem.

When your pup is 4-5 months old watch it closely, and if you detect it using a lot of eye on sheep, don't let it 'stick up' (become mesmerised by the situation), keep it moving and keep the mob moving so the pup will be forced to cover the mob. Only use sticky eyed dogs on large mobs — don't use them on three or four sheep.

It is most important to break the dog's concentration and get its attention. Cracking a whip or throwing something near the dog normally does the trick, although I have seen some dogs so engrossed in holding a sheep that, a shotgun fired in the air was needed to break their concentration to get them off. When their concentration is broken immediately give a directional command and attempt to keep them moving.

*Consummate
control*

Most hard eyed dogs are very strong and I have found them very useful. When working cattle, they tend to free up and because of their strength usually will nose bite effectively. Sticky eyed dogs usually won't bark, it is almost impossible to make them bark at anything, a highly desirable attribute for dogs working cattle.

The loose eyed dogs are also not as effective — they tend to lack the ability to balance and hold a mob. These dogs are normally hyperactive, and will chase a mob barking madly. They usually lack intelligence and are difficult to train because they can't be kept calm. These dogs are usually the result of poor breeding and unfortunately are quite common on farms throughout Australia.

The best way to handle this type of dog is to get control. Reinforce its basic obedience training so it can be stopped when it becomes over excited.

Never use a loose eyed dog on cattle as cattle will tend to excite them even more. This type of dog also tends to stir cattle up to a stage where they will be difficult to handle even with a good dog. These dogs are best kept for yard work, because they can be kept in check more easily and problems such as too much bark and over-enthusiasm can be turned into assets. Most dogs can be taught to bark, so instead of having a problem-ridden paddock dog, you can develop a more-than-handy yard dog.

Hard or loose eyed dogs are undesirable and breeders should be discouraged from continuing their line for both types are hard to train as all-round workers, which really is the purpose they were bred for in the first place. However neither is absolutely useless and with adaptation, and with the application of a little lateral thinking by the handler both types of dog can be quite alright performing tasks to which they are more suited.

Well bred working dogs will have the correct amount of eye and will only use it when they need to hold or balance the stock. It will not be evident when working large mobs or working in the yards.

Driving

Most stockmen when moving sheep or cattle from one point to another work their stock from behind and hunt them away. The Australian Cattle Dog and the New Zealand Huntaway perform this task quite easily as they are principally driving dogs; the Kelpie and the Border Collie are principally heading dogs and need more intensive training to learn to drive.

Heading dogs should not be taught to drive until they are 10-12 months old and casting and drawing sheep to the handler very confidently. The most common complaint I hear from handlers when driving a mob is that 'the rotten dog keeps getting in the way and wants to bring the mob back to me all the time'. If this happens take time to have a look at the handler's position relative to the mob and you will realise the dog is working well but 99 per cent of the time the handler is in the wrong position.

To teach your dog to drive the following method has always proven very effective:

- Get a quiet mob of 10–20 sheep.
- Walk toward the sheep with the dog behind you.
- Send your dog to the right (in an anti-clockwise direction).
- As it approaches the right wing or the 4 o'clock position STOP IT
- The dog does not need to sit, however if you are having trouble with the dog it may be necessary to sit it.

Give your dog a directional cast

*Put your dog at
4 o'clock position and send
it back to 8 o'clock position*

*It is preferable your dog runs
behind the handler*

69

- Call the dog back to you and send it left (in a clockwise direction).
- Stop it at the 8 o'clock position.
- While the dog is performing this task you will find the mob starting to move away from you and the dog. If you follow the mob and repeat the procedure a few times the dog soon learns to drive.

Once again this sounds very easy, however you are likely to encounter a few problems:

- The dog keeps running to the head of the mob.

If the dog keeps running to the head of the mob, it indicates a lack of understanding on the dog's behalf, bearing in mind they are instinctively a heading dog and until now all their training has been designed at getting them to bring the mob back to the handler. Reinforcement of the basic obedience training is all important and perseverance will eventually have your dog stopping on the wings of the mob.

- The dog gets confused when it is stopped

In situations with soft or timid dogs confusion occurs when they are stopped on what they believe is their cast. Here the handler needs to reinforce the action with verbal praise when the dog stops. Give your dog plenty of praise when it comes past you and again when it goes the other way. Good results will occur when teaching to drive this way.

I will stress again not to attempt to teach your dog to drive before it is 5–7 months old. This can have devastating effects on its ability to cast and in some cases will stop it from casting all together.

Controlling Stock

Shepherding refers to a dog's ability to control stock, and once again is instinctive. When watching a dog work you can see how it covers the total mob but has the ability to drive and control the spread of the mob at the same time. A dog with a natural shepherding ability needs no direction when stock start to stray, but will move steadily out to the wing and turn them back into the mob and return to the back of the mob and keep driving. These dogs also possess great natural anticipation, the uncanny ability to know where and when stock are going to move and they seem to be there waiting for the stock when it happens.

A dog with little shepherding instinct normally has to be directed to perform this task, even though with time most good dogs will develop this ability to varying degrees. As with most inherited traits not a lot can be done with dogs lacking ability. However if the handler directs the dog to the correct spot and encourages it to watch the mob it may develop some of the anticipation required for shepherding.

This ability is more prevalent on well bred dogs than in the thrown-together lines. It is often referred to by people when they say this dog is a 'natural' — 'I didn't have to teach it anything'.

Working dogs from a truck or motor bike

The practice of working dogs from a truck or bike is common throughout Australia. Problems do arise from this practice when a handler attempts to train a young dog from the cab of a truck or the saddle of a motor bike. All training must be done on foot. If you are sitting in the cab of a truck, a young dog will become confused with voice commands when it can't see you.

Motor bikes cause different problems because the rider isn't using the dog at all. Normally the rider does the casting on the bike and the dog chases the bike or sits on the back and doesn't learn to cast at all. Remember there's no point in having a dog and barking yourself. Be kind to your dog and use it for what it was bred for, a working dog not a pillion passenger. Make a point of stopping and parking the bike a suitable distance from the mob then cast your dog, let it bring the sheep back to you and then help the dog drive the sheep to the yards.

Working stock with more than one dog

Australian stockmen frequently work stock using more than one dog, however many don't do it very well. A number of factors contribute to this, the main problem is usually that the dogs are under-trained. There is absolutely no point in working two dogs if both are in the same position on the mob because the net result is two dogs doing less work than one dog would do alone.

However, if you are handling two well trained dogs, they can be correctly positioned and controlled and the job becomes easier for the stock, the dog and especially the handler. The ideal situation when working two dogs is to have one working the wing, while the other is driving from the rear, or on a very large mob (more than 200 cattle or 700 sheep) one dog on each wing while the handler follows the mob keeping it moving. Directional commands come into their own when working more than one dog.

Never attempt to train two dogs together. The only thing that you will achieve is total confusion.

Driving with ease

7

Yard Work

The movement of stock through yards should be done as efficiently and effectively as possible. Over the years I have witnessed some mournful attempts at working sheep through yards. Although, finally, they achieve their main objective the stock are left stressed. It is essential to stress sheep as little as possible, especially ewes in lamb as it can lead to problems such as pregnancy toxaemia and wool tenderness.

Many people believe yard work is a totally separate thing to paddock work. However, I believe yard work to be an extension of paddock work.

I would not recommend introducing your dog to yard work until it is at least 9–10 months old. Yard work requires the dog to be both physically and mentally mature. An indication of the right time to start yard training is when your dog is confidently and competently performing paddock work.

Many times I've seen dogs whose handlers believe them to be good yard dogs, barking madly at the back of a big mob, with sheep at the rear stacked on top of each other ten deep and the sheep at the head standing still and totally oblivious to the commotion at the rear. Watching performances like this accentuates the need for some additional specialised training in the yards.

Not all dogs will make good yard dogs, although a keen worker can be trained to be more than handy in the yard. If your dog is a soft and gentle paddock worker and is obviously timid when in close proximity to stock it would be better to allow this dog to do what it does best and work it only in the paddock. If pushed into yard work a dog of this nature will most probably become a biter, an undesirable characteristic, and could also ruin its paddock working ability.

I have mentioned earlier that paddock work is mainly instinctive, the handler really only develops and controls the dog's natural instincts. However yard work requires the dog to actually learn new skills and do things totally foreign to instinctive behaviour.

Backing

The major purpose of all yard work is to move the stock efficiently and effectively through the yards. Keeping this in mind the ability to correctly position the dog is essential. In all cases it is highly desirable to have a dog with the ability to move stock efficiently through a race into a catching pen or onto a truck. Three methods come to mind immediately, two good and the other not-so-good.

The not-so-good method and the most common one used is to have the dog run back and forward outside the fence barking its head off and giving the occasional bite. This method is not very efficient and usually indicates an under-trained dog.

The second method (running the race) has the dog running outside the fence until it reaches two or three sheep from the head of the race,

Confident backing —the essence of yard work

jumping the fence into the sheep and running back along the ground through the mob. This is a useful method but has its limitations especially if stock is being loaded onto a truck or into catching pens.

Backing is the third and best method. Backing means the dog runs along the sheep's backs toward the head of the mob and jumps onto the ground two or three sheep from the end and then run back through the mob along the ground. This is a very effective method and is ideal for all yard work.

Teaching your dog to back

Earlier in its training your dog will have learnt to perform the **up** command. This could also be **jump up** or **get up**.

To teach your dog to back, fill a drenching race tightly with woolly sheep (woolly sheep make it easier for the dog to grip their backs). Standing as close to the sheep as possible give the **jump up** or **get up** command. There may be some hesitation from the dog, if so lift it up and place it on the sheep's back. Encourage and reassure the dog. When the dog appears comfortable attach a short lead (1 metre only) to the dog, give the come command and slowly lead it along the sheep's backs. When it comes readily give it plenty of praise.

The next step is to lead the dog along the length of the race giving the command **go back**. When you reach the end of the race lift the dog onto the ground. At this stage don't attempt to teach your dog to run back through the sheep as it won't yet have gained enough confidence.

As you know patience and perseverance are the keys to success. The above training will not bring instant success. The dog will probably refuse to jump up and when you place it on the sheep's backs it may sit relatively calmly but at the first movement it will jump straight off and head for the furthermost corner of the yard. Once you manage to retrieve the animal and repeat the procedure ten or so more times and you finally get the dog to the other end of the race you will both be exhausted and very glad the lesson is over.

Guidelines to remember:

- When the handler places the dog on the sheep's back speak calmly and reassure the dog with plenty of praise.
- Don't let the dog jump off the sheep's back. If it does, show your disapproval strongly (do not physically punish) and immediately place the dog back on the sheep.
- When teaching a dog to lead along sheep's backs, do it in small stages, about a metre at a time, compliment and pat the dog at each step it takes.

Get in there!

- Lifting the dog from on top of the sheep when the lesson is finished reinforces the fact that it is not to jump off the sheep's back.

The length of the first lessons should be kept short. If your dog is very uncomfortable with being on the sheep's backs it is advisable to end the lesson as soon as the dog appears to be gaining confidence just standing on the sheep's backs. Repeat the lesson the next day and gradually lengthen the duration until the dog is performing the task comfortably.

If your dog is quite happy on their backs and performs well, repeat the entire procedure. When the dog tires stop the lesson by lifting it off the sheep and continue the training on another day.

When your dog is running along the sheep's backs with confidence, it is time to remove the lead and attempt the procedure without it. While it is standing on the sheep's backs give the **go back** command and walk next to it as though the lead is still on it. If the dog is ready to perform the task without the lead try sending it along the sheep's backs while standing at the end of the race. When it does this correctly the dog is ready to be taught to come back through the mob.

Teaching the dog to come back through the mob

This is an essential part of yard work. If your dog will run back through a mob in a race you will achieve your aim of moving the mob through the race.

Backing is the means of correctly positioning the dog and you should not attempt to teach the dog to come back through the mob until your dog is backing confidently. This is important, as a dog that is not ready for this stage might get squeezed between the stock and could get hurt.

- Attach the short lead to the dog once again and **back** it the length of the race with you walking beside it two or three sheep from the end of the race.
- Stop it, then call it to you and at the same time encourage it to jump to the ground between the sheep.
- Lead it through the sheep, making sure it doesn't get hurt. This lesson is usually quickly learnt and results appear within a couple of lessons.

Coming back through the race with strength and confidence

Some points to remember are:

- Don't let your dog jump out of the yard.
- At first lead your dog all the way through the mob to the rear.
- Keep an eye out for biting. If your dog is going to bite, this is the stage of the training it is likely to do so.

Running the race

With the dog on a short lead and in with the sheep at the rear of the mob, call it to you from outside the fence.

- When it comes to you lead it to the head of the race about two or three sheep from the front.
- Give the command for it to jump into the race and lead it through the sheep making sure it doesn't get hurt.

When the dog is performing this task well while on the lead, try it without the lead.

- Call the dog to the position you want it to enter the race from.
- Give the command to enter the race.

Alternative method of running the race

- Walk along the outside of the race next to the dog as it moves through the race.
- The dog should perform this stage quite readily.

  The final stage in training is to

- Send the dog to the head of the race
- Give the command for it to enter the race and call it back to you at the rear of the mob.

8

Working Cattle

C attle are much easier to handle when worked quietly and calmly. Unfortunately some stockmen here and overseas haven't grasped this fact. I have witnessed some very poor efforts at mustering, reminiscent of an old Western, with stockmen seeing a mob and charging madly at it yelling and whistling enthusiastically. Their dogs see all this commotion and join in, happily helping the stockmen drive the mob through the fence and giving each beast a bite on the way through. The last thing seen is the mob, the dog and the fence disappearing over the horizon and into the neighbour's place. It really doesn't have to be like this. If the mob is worked quietly and calmly, with your dog under control, cattle can be yarded quickly and efficiently without any stress placed on the cattle or the stockmen.

Force and control — essential for cattle work

In earlier chapters I have concentrated mainly on methods used to train dogs for sheep work. These methods work equally as well with dogs required to work cattle, however there are a few significant differences. Pup selection is most important when looking for a cattle worker. The pup should be selected from bloodlines bred specifically for cattle work and preferably from proven bloodlines that deliver good cattle workers.

A dog that is a good cattle worker must have the strength to head and heel the bigger animals. All breeds of sheepdog will work cattle to a degree, however they will work them as though they are sheep. Normally there's not a lot wrong with that but there are two important differences between sheepdogs and cattle dogs.

Most sheepdogs aren't strong enough on the head of cattle. When a dog is weak on the head it will cast to the head of a mob and then will either bark, turn tail or run away, mainly because it is frightened. This is not a good trait for a sheepdog either, but the sheer size and aggression of cattle will sometimes bring out this trait in otherwise excellent sheepdogs.

The other major difference is a cattle worker's ability to heel (bite low on the hocks of cattle). Dogs specifically bred for sheep work will not possess this ability, in fact they are strongly discouraged from biting. A dog that specifically works cattle needs to be stronger and more aggressive when required to block and move cattle. A cattle worker also needs to be able to cast to the head or lead of a mob and block with force. This is usually achieved by nosing or biting the nose of the lead beast to block it, then covering the mob and holding them. A dog that won't cast to the head will never make a good cattle working dog. A dog's ability to heel is also an advantage when getting cattle moving in the direction the handler requires. After the dog has blocked the head it is also handy if it is able to heel a beast when it turns back into the mob.

Strong debate exists between breeders and trainers whether it is better to start a pup that will be working predominantly with cattle, on sheep or on cattle. My recommendation is to start the pup working sheep because it is easier to keep the pup keen and under tighter control when it's young then change it to cattle when it's 8-10 months old.

However if you have no access to sheep, there are methods to be used when starting pups on cattle.

Basic obedience should be taught the same as for sheep. I stress again it is a most important phase of a dog's training. A major problem I have noted from my experiences is that dogs used to work cattle are not controlled enough. The basics **come**, **stop**, **sit**, **down**, **stay**, are still vital in the training of cattle dogs and must be enforced.

Introduction to Cattle

Begin familiarising your pup with cattle when it is 4–5 months old, making sure these cattle are quiet and not aggressive toward the pup. Don't put young pups with cows and calves. A cranky old cow chasing a pup down the paddock will put its development back months (one of the reasons I recommend starting pups with sheep). Let the pup watch weaners and develop some eye and let them watch other dogs working.

When your dog is responding well to the basic obedience commands and has been well familiarised with the cattle at about 8–9 months, it is time to start it on a mob of 12–20 weaners. This is also a perfect opportunity to quieten young weaners and get them used to being worked by dogs, and this exercise will not only help your pup but will mean quieter control of your cattle in the years to come. A yard 30 metres by 15 metres is ideal to start this training. An old experienced dog is also very useful during this stage of training especially if the young dog is lacking confidence. This is one of the major differences between training sheepdogs as opposed to cattle dogs, as quite often a second dog is needed to be worked with the young dog. Care must be taken however not to let the young dog become dependant on the older dog, so work the young dog by itself when it develops enough confidence.

Take the dog into the yard with you and let it go. Watch what it does, and if it wants to run to the head of the mob and block, encourage it to do so. Praise the dog when it does this. When it is running to the head and blocking it is time to move to the next stage, casting.

Casting

Try to get the weaners at the opposite end of the yard to yourself and preferably moving away. Sit your dog and position yourself between the mob and the dog, then give the command you use to cast and encourage the dog to head the mob. When it heads the mob and turns the lead beast in on the mob give the **sit** command. When the dog obeys give plenty of praise. The handler's *position* is most important during training. Positioning yourself between the mob and the dog gives you the opportunity to push the dog out if it casts too close. A dog that close casts will come onto the mob too flat and will be on the lead beast's shoulder rather than being in front, giving the dog less initial control.

After casting your dog and allowing it to block and turn the weaners encourage it to draw the weaners toward you. If you have a strong heading dog it will be inclined to run around to the head of the mob. If so stop it and send it behind. If the dog is very strong on the head it may be necessary for the handler to position himself close to the mob to block the dog then send it behind. This action is in effect the first stage of a drive and should not be attempted until the dog is casting, heading and blocking confidently.

Casting anti-clockwise

Casting clockwise

The dog should be cast from behind the handler

Ensure your dog stays wide

Stop at the end of the cast. Control is all important

86

Under control

*Blocking —
essential for
working cattle*

87

Teaching your dog to block

If your weaners are in one yard, open the gate into a larger yard. With your dog under control and by your side walk around the cattle and let them move into the other yard. On seeing the cattle getting away the dog should instinctively want to head them. Give the dog the command you use to cast and allow it to block the head of the cattle with force. When the dog and the cattle have settled, allow the cattle back into the other yard and repeat the cast and block procedure.

When your dog is performing these tasks confidently and is under good control try it with weaners in a larger paddock (1–2 acres is ideal). Repeat the procedure for cast, block and draw. If the dog has learnt its lessons well the handler will have no trouble. However it is more likely that the dog will refuse to obey the control commands. If the dog completely ignores you it obviously was not ready for work in the larger yard, and you will have to reinforce the control commands again in the smaller yard. Remember, basic obedience is the basis of all training, you, the handler, must be in control.

Force

It is essential for a good cattle dog to have force. Some dogs are born with lots of guts and the ability to force, and these dogs usually need to be encouraged to be calmer, rather than get excited. A softer dog however, with lesser ability to force can be encouraged with a few methods.

Set up situations where little can go wrong and work the dog on quiet weaners to build its confidence. If the dog has trouble and the cattle refuse to move, step in and give it a hand, encouraging the dog to move in and force.

Sometimes it is of great benefit to excite the dog to the point when it will rush in forcing and biting, however the dog needs to be well controlled so it can be stopped when necessary. Another alternative is to work the dog with an older experienced controllable dog to develop its confidence. Be very careful the young dog doesn't start to rely on the older dog.

Bark

A barking dog stirs cattle up and makes them harder to control. Cattle breeders don't always agree that this is a fault, and there are instances where a bark is an asset, such as when cattle are running hilly scrub

Force — strength and determination

Excuse me!

country and the dog has been cast blind into heavy timber. In this case a dog that barks when it finds the cattle is a big advantage as it means a clean muster.

Direction

Having a dog that will move as directed is as important with cattle as with sheep. A lot of good dogs aren't taught directions but they usually have great powers of anticipation and are therefore usually in the right place. If you need your dog to respond to direction commands you will have to teach it. To teach direction use the same method as for sheep, however I would recommend initial training be carried out away from the stock when the dog is 7–8 months old. You can use either a training stick or a rotating pole. These methods are fully described in Chapter 6.

After your dog has learnt each direction and is 9–10 months old it will be mature enough to be tried on cattle. Once again get a mob of weaners in the small yard, position yourself close to the mob and give the **away** command for anti-clockwise and allow the dog to run in an anti-clockwise direction to the head of the mob. Give the **by** command for clock-

Teaching direction using natural balance instincts

Having a dog that will move as directed is most important

wise direction and the dog should return to where it started from. Repeat the process for clockwise direction onto the other side of the mob, again stopping it at the head of the mob. After it is performing direction well in the yard it is time to start teaching the dog to drive.

Driving

Working dogs must be able to drive because most stock work is performed from behind the mob. Natural driving dogs like the Australian Cattle Dog are easily taught as the action is mainly instinctive, however with natural heading dogs like the Border Collie and the Kelpie the task can be much more difficult.

I have enjoyed good success using the following method:

- Using a small mob of free walking quiet weaners in a 1–2 acre paddock, walk with your dog toward the mob.
- Stop about 20 metres from the mob and send your dog with the clockwise command to the 8 o'clock position.
- Stop the dog and give the anti-clockwise command.
- Once again stop the dog, this time at the 4 o'clock position.
- While giving these commands move toward the mob and the dog will soon learn to drive the mob away from you.

91

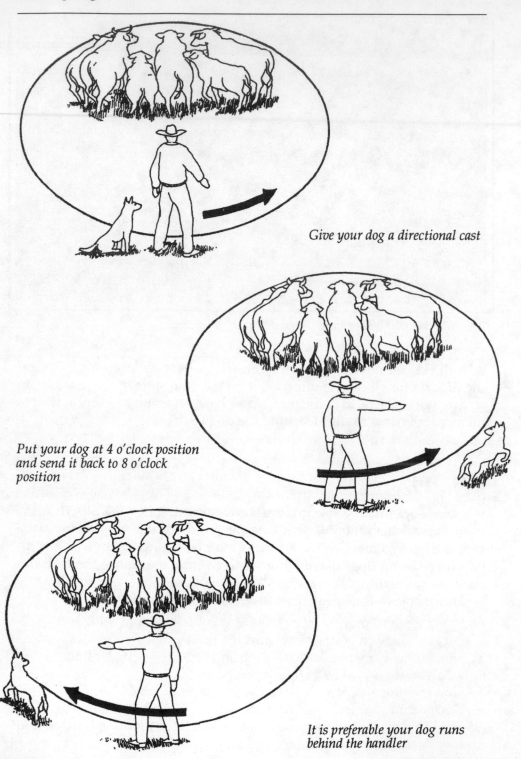

Give your dog a directional cast

Put your dog at 4 o'clock position and send it back to 8 o'clock position

It is preferable your dog runs behind the handler

Turn back dog used for training cutting horses

Initially if you drive the cattle along a fence line it will make the task much easier for the dog. Position yourself with your dog beside you, behind the cattle. When the cattle start to move away cast your dog to the rear wing and stop it. Call the dog back to you, then repeat the procedure. This method is quite successful as the dog only has to watch one wing. Make sure when using this method that you work the cattle in both directions up and down the fence line to ensure the dog learns to cast to both wings.

When the dog is performing this task without too many problems it is ready to be tested on a larger mob of older cattle. However until the dog is working with a great deal of confidence and is both physically and mentally mature it should not be used on aggressive cattle.

Training Cattle Dogs on Sheep

As mentioned earlier in this section I prefer to train cattle working dogs on sheep first and introduce them to cattle at a later stage. The method I use should be used on dogs that are to work cattle only, not those who will be working cattle *and* sheep. Dogs used solely for cattle work are encouraged to show more aggression and should not be discouraged from biting.

Basic obedience through to casting and driving should be performed on sheep as described earlier in the book. Except when casting don't push the dog as wide on its cast — it shouldn't be working too wide from a mob — and allow it to work a little more aggressively when it is lifting and drawing. A dog needs to show a fair amount of force to lift and draw a mob of cattle.

Care should be taken to ensure good control is maintained over your dog at all times. When your dog is working well on sheep it can be introduced to weaner cattle in the small yard at about 10 months. Most of the time the transition will be quick and painless.

If you require your dog to work both sheep and cattle a compromise must be reached with regard to the amount of aggression that will be tolerated. However many well bred dogs can adjust their aggression and cast to the type of stock they are working, and as the dog matures this initiative can be encouraged.

Big versus Small

A popular misconception exists about the ideal physical size required for dogs working cattle. A lot of people think they need large dogs, but in most cases large framed heavy dogs are just as useless as small yapping dogs because they lack the agility that is essential in a good working dog.

A dog working cattle needs to be reasonably fast moving, so it can get to the head quickly and block as well as cover the mob. Currently in the United States of America the quest for bigger and bigger working dogs is reaching a point where certain breeds of working dogs may be affected to their detriment. Remember the old adage 'Its not the size of the dog in the fight, its the size of the fight in the dog.' Australian working dogs have evolved over many years and have been refined to possess the qualities required to handle stock efficiently and effectively. Breeding to increase size only can only lead to a regression in a dog's working ability.

9

Behavioural Problems

9

No matter how well bred they are many dogs do develop behavioural problems.

One of the worst is sheep biting. This, if allowed to continue, can develop to become so serious that eventually the dog has to be destroyed.

Other problems are unnecessary barking, chasing cars and chasing and/or killing chooks, fighting other dogs and jumping on people.

Most of these problems are easily fixed, but as in most cases it is essential to stop them before they start. I have used some different methods to curb the problems from time to time; usually the most successful have been the simplest.

Biting

If you own a sheepdog that bites even to a small degree it has to be stopped. Don't wait until it is sending sheep lame or killing them before you act. Any biting is still harming the animals by either pulling out wool or placing undue stress on the animal.

A sheepdog that bites has to be stopped. Fitting a muzzle stops the stock from being bitten but doesn't stop the dog from trying to bite. If you have allowed your dog to bite from the time it was a pup you may have left your run a bit late — early prevention is the best cure.

Never work a young sheepdog with a biting dog. It is the quickest way to teach the pup bad habits.

If your pup bites, it must be disciplined immediately, preferably while it still has the sheep in its mouth. Hitting the dog and scolding at the same time will often stop the biting. Timing is the vital ingredient. The dog has to know why it is being disciplined.

It is worth noting that a too aggressive approach by the handler can be misunderstood by the dog. It may think it is in trouble for working the sheep rather than biting and consequently may become hesitant to work.

Not all biting should be looked on as being bad, especially in cattle

96

dogs, but a dog that bites a sheep on the flank or hind leg and won't let go, should be dealt with firmly and quickly. A dog however, that will stand its ground and bite a charging sheep on the nose, should be viewed differently. In most cases it is the sign of a good dog. A dog that will do this shows great resolve; it has limited methods of defence, it can stand and bark, it can run away or it can stand and fight. Any dog that will take on something three times its size, while it is charging straight toward the dog with the intent to kill, will prove to be a very handy dog. Most *people* would probably turn and run.

In short, a dog should never be punished for defending itself, but should be firmly punished for indiscriminate biting.

Barking

The constant sound of dogs barking can severely test the patience of everybody. It very quickly irritates and annoys anyone within earshot. Although having a dog that barks can be a benefit, the importance of having a working dog that barks is highly overrated. Working a dog that is silent, that will back, and come back through the mob will move far more sheep than one that barks non-stop at the rear of a mob. The perfect situation is to have a dog that will bark on command and go about the rest of its business quietly.

Dogs usually bark for one or two reasons: they are frightened or they are excited. The response of barking at either of these emotions is a highly undesirable trait in working dogs. The ideal dog remains calm and in control of the task it is performing. This applies to both yard and paddock work.

A frightened dog creates all sorts of problems for itself and its handler. Usually it will bark non-stop at a safe distance from a mob of sheep and the sheep soon learn to ignore the noise. Frightened dogs tend to be the biters, they bite out of fear when put on the spot by one or two sheep or if a mob gets too close to them. These dogs are usually better kept for paddock work where they can keep their distance from a mob.

An excitable dog is marginally better as it is normally less inclined to bite, however this type of dog is usually not very intelligent, thus difficult to train. Anyone who has worked long hours in the yards with this type of dog knows how annoying non-stop barking becomes and after a while the sheep also get used to the noise. The dog remains annoying but becomes ineffective.

Dogs allowed to bark when chained up or when on the truck are also a real nuisance and should be stopped for the sake of everyone's sanity.

Dogs seem to start barking when they are chained, and at 1 or 2 o'clock in the morning. They manage to wake the entire household forcing some member of the family, usually the handler, to get dressed and go down to the dog yards to discipline the offending animal.

I have found it much better to train the dog during daylight hours rather than in the middle of the night.

One method I use is to set up a lawn sprinkler near the dog. Attach it to a long hose and when the dog starts to bark give the **sit down** command or **shut up** command and turn the hose on. The dog soon associates the command with getting wet and usually stops barking.

Another method sworn to be absolutely fail-safe by a local dog breeder is the 'dunny roof' method.

- Chain the dog to an old tin dunny.
- Collect a bucket full of rocks.
- Get out of sight of the dog.
- Wait till the dog barks.

- When the dog barks throw a rock onto the dunny roof.
- Before the rock hits the roof give the **shut up** command.

The noise made by the rock hitting the roof will stop anything from barking.

As silly as this method sounds it actually works. The only problem I have found with it is the shortage of old tin dunnies. However, if you tie your dog in a shed or to a galvanised iron fence it will have the same effect.

This same dog breeder has given me many hilarious methods of solving training problems. I thank him kindly for each one. This is a good time to re-affirm a point, a good sense of humour is just about essential if you are training dogs.

If your dog barks while travelling on the back of the truck you can try abusing it out the window. If that doesn't work attach a light piece of rope to its collar and put the other end through the window. When the dog starts to bark jerk the rope firmly while giving the **shut up** command.

A friend of mine had a problem with his Australian Cattle Dog barking on the back of his truck. He waterproofed a speaker and hooked it into the intercom connection on his CB radio, mounted the speaker near the dog's ear and as soon as the dog barked he turned up the volume as high as possible and told the dog to shut up. He had a very startled dog but after a little perseverance the problem was cured.

This basic method could be used in a lot of different situations.

Chasing Cars

If your dog likes to chase cars, you have to stop it and the sooner the better. If your dog is responding well to its basic obedience training, command the dog to **sit** as soon as it starts chasing the car should break its desire reasonably quickly. If however, this method doesn't work, you may have to set up a situation in order to stop it once and for all. One method is to stand on the tray top of the truck armed with a stock whip and when the dog starts to chase the truck crack the stock whip and reinforce the **sit** command. This usually does the trick.

Killing Fowl

Wandering down to the fowl yard to find dead chooks all over the place does not endear your dog to you, no matter how good a worker it is.

By introducing a 3–4 month old pup to fowls in a set-up situation you can actually stop the problem before it starts. Let your pup loose in the fowl yard, and as soon as the chooks move the pup will become interested and will start to work them. If it continues to work them everything's alright. Fowls however tend to excite dogs and bring out their natural killing instinct. If your pup rushes in to bite one grab the pup immediately and discipline it firmly. Then put the pup back down and encourage it to work the fowls. Continue this method until the pup is happy just to work them rather than kill them. I have seen older dogs, which have been taught by this method, working chooks into a corner and then lying on them to hold them down, with no harm to the fowl, in fact without even losing a feather.

If a dog is allowed to keep killing fowls it is very difficult to stop, and unfortunately many good dogs have been put down because they wouldn't stop. Sometimes I think a well constructed chook run is a better alternative.

Fighters

Owning a dog that wants to fight every time it sees another dog is at best awkward and can cause considerable problems for the handler. This problem should be addressed as soon as it begins. Normally stern discipline to the aggressor will cure the problem, however if you have a dog or have acquired a dog that is incorrigible you will need to take more drastic action.

You need to use another dog and another handler. If another handler is unavailable put the other dog on a short chain and tie it to a post or a fence. Grab yourself a short piece of poly pipe and put it in your back pocket. Put a short lead or chain on the problem dog and position yourself with the dog about 50 metres from the dog it wants to attack.

Walk slowly toward the other dog and at the first lunge, growl or sign of aggression from your dog smack your dog across the muzzle with the poly pipe. Continue walking toward the other dog, forcing the fighter toward it. After being belted the fighter will not be all that interested in the other dog and will probably try to turn the other way, but make sure you force it towards the other dog. If it shows any sign of aggression belt it again, and continue to force it onto the other dog. You will be surprised how quickly the fighter will learn not to attack. Persistence with this method will cure most aggressive dogs, but should only be used as a last resort. The majority of fighting problems are normally sorted out when your dog is 15–16 months old and first displays signs of wanting to fight. Stern discipline at that age nips the problem in the bud. I repeat, *this heavy handed discipline should only be used as a last resort.* Some people may consider it cruel, however when you look at the alternatives it may not be cruel at all.

Jumping on People

Murphy's Law states that the propensity for your dog to charge at and jump all over you is directly proportional to your mode of dress and the amount of time you have left before going out for the evening.

Friendly pups frequently bound across the yard and launch themselves at you in a show of affection. This action is a normal part of the bonding process, however it is best stopped before it becomes a habit. More often than not this form of jumping can be stopped during basic obedience training by giving the sit command before your dog leaps all over you.

If however your dog persists in jumping all over you more drastic measures are required. I have found the knee a most effective deterrent. Timing is all important. When your dog comes bounding toward you and is about to jump, raise your knee and if your timing is right the dog should bounce off you and hit the ground a short distance away. Normally the dog learns very quickly not to jump on you or anyone else. Be careful to ensure the timing is right, because if it isn't the dog will think it has found a new game and will continue jumping.

*With correct training and attention to good behaviour dogs can be taught almost anything. Backing **cattle** is not recommended because it can be dangerous to the dog*

10

Additional Skills Look, Lead Dog, Dog Trials

10

A dog that will look for stock is exceptionally useful. Lead dogs that will drive stock slowly and prevent the mob from moving too quickly are also a godsend to drovers and graziers.

These two special skills will not be present in all dogs and some may never learn them. However, your dog may show an aptitude for one or both. If so, persevere. It will be well worthwhile.

One way of learning about the amazing qualities in our Australian working dogs is to see them perform with their handlers at dog trials.

Look

The ideal time to begin training your dog to look for stock is when it is 14–16 months old. Some dogs possess a wide searching casting ability and actually look for stock during their cast, using their senses of sight, sound and smell. Some dogs, although possessing a good cast, don't have a searching cast. Most can be taught to look for stock.

If while casting your dog misses some stock immediately stop it and walk toward the stock it has missed and give the command **look**. By incorporating a hand signal, most dogs will learn to search for the stock they have missed on hearing the command and will run in the direction the your are pointing.

Lead Dog

Many people mistakenly believe a lead dog to be a heading dog. This belief although not totally unfounded is not true. A true lead dog is used by drovers and graziers who want to drive stock slowly and prevent the head of the mob moving too quickly. Lead dogs come into their own when moving a mob containing weak or sick sheep, they slow the mobs down more to the pace of the weak ones. A true lead dog can be cast to the head of the mob, but instead of bringing the stock back to the handler, it will give ground when the stock approach, slowing the mob's progress but not blocking and turning.

On first impression you may think a lead dog is a weaker frightened heading dog. In some cases this is true, but the good ones have the ability to lift and draw on request. I have found lead dogs are born, not made; most perform instinctively and good ones are quite rare. This places them in high demand and most drovers would give the shirts off their backs to own one.

Dog Trials

Many farmers and dog handlers are unjustifiably critical of dog trials. Both three sheep trials and yard trials are an essential element in the progressive improvement of working dogs. Although of course not everyone requires a dog that will carefully move three sheep through an obstacle course, the quality of breeding required to develop these dogs to

Finesse in training

this level — to perform beautiful casts, lift and draw, footwork, cover and eye — are essential for dogs that are used in the pastoral industry throughout Australia.

These attributes are often exaggerated in trial dogs, however the skill must be retained in our working dogs, and anyone interested in breeding top quality working dogs should pay heed to them. Anybody that has cast a dog into hilly country to recover a few stragglers and watched as the old dog quietly moves these sheep down the hill, through the gate and back to the rest of the mob, should thank the old masters (most of whom were trial men) who bred these wonderful qualities into our working dogs.

The best place to learn about working dogs and handling working dogs is on a trial ground, talking to the top men and women in the country. If the opportunity arises and you get the chance to pencil (score) for one of the better judges you should take it. You will be amazed how much knowledge there is to pick up. If you are keen to polish up some of your handling skills enter some of these trials, as the trial ground is the place to learn.

Yard dog trials have increased in popularity over the last few years. There is plenty of action and they are good to watch. I believe it is easier to compete in yard trials because the handler may assist the dog and the finesse of three sheep trials is not required. Although top level competitors in this sport require great skill in dog handling, I do believe it is a good place for beginners to watch and learn and benefit from the hours of training put in by the better handlers.

11

Caring for your Working Dog

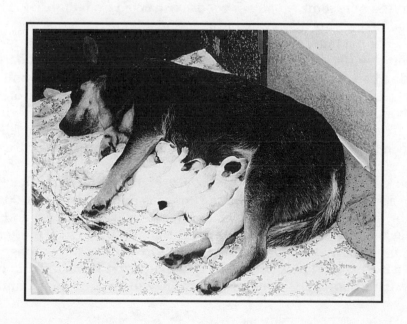

11

In my travels around Australia and the United States attending field days, performing working dog exhibitions and visiting properties, I have seen at first-hand the poor conditions in which a lot of working dogs spend their non-working life. It makes me wonder why some people have dogs if they won't look after them properly. A little attention to detail might just make a few more potentially good dogs even better.

Housing

Some dogs aren't really housed at all, but just tied up under a tree or chained to a 44 gallon drum that gets boiling hot in summer and freezing cold in winter.

Properly constructed kennels are not very expensive and are relatively easy to construct. I have seen some very good kennels made of brick and galvanised iron, some equally as good constructed from timber. Pre-constructed kennels are also available at reasonable cost.

A better but more expensive alternative to kennels is a properly constructed dog run. These are an absolute necessity if you are embarking on a dog breeding program, or if you own a bitch and don't want any unexpected pregnancies. These runs consist of a kennel area made of brick or galvanised iron with a three metre run enclosed with chain wire or mesh. The run area should be at least 1.5 metres wide. The run itself can be concreted or left as dirt, although concreted runs are easier to keep clean.

The kennel pictured opposite, above, is one of the best constructions I have seen. The total cost of these runs was approximately $300 and most of the materials used in the construction were off-cuts and bits and pieces found lying about the property. The total length of the kennel is 6m, each individual compartment is 1.2 m wide and 2 m long. The compartments are lined with fibro and insulation is installed in each cavity. The flooring is constructed from woven wire mesh and an

The Taj Mahal

The People's Palace

automatic watering system has been incorporated. Its construction 1 m above the ground ensures the kennels are cool in summer and stay clean and dry. A point worth noting is that the small size of the runs requires the dog to be let out every day.

An improvement to these runs would be to incorporate a two metre high fence around the kennel 20 m × 20 m in perimeter to allow the dogs to have a run, either all together or individually, then they could be left out of their kennels for an extended period of time.

Bedding

Under no circumstance let your dog sleep on the ground. A bedding board above the ground is ideal.

The issue of whether or not to use bedding is an interesting one. I have seen dogs chained to well constructed kennels with very good bedding. But what does the dog do? It digs a hole in the dirt and sleeps in the rain. Kelpies are notorious for this.

In spite of this if a dog is housed in a concrete run the kennel area must contain some kind of bedding to raise them above the floor. A bed board measuring 100 × 60 cm and constructed from floorboards with some bedding on it does the job. The bedding material can be anything warm, an old wheat bag, a blanket or any cotton material.

The bedding must be replaced regularly to prevent a build-up of fleas and worm eggs.

Position of the Kennel

When selecting a position for your kennels think of a few basic considerations:

- shade
- cleanliness
- relationship to homestead.

A great deal of thought should be given to the positioning of the kennel especially if it is going to be a permanent construction. It is essential to think about shade, proximity to the homestead, proximity to the yards. Dogs have their own distinctive odour, they bark, kennels attract vermin, so when you are considering constructing kennels keep these points in mind.

Cleanliness

Dog runs and kennels must be kept clean, to prevent worm infestation, fleas and diseases. Dog runs with concrete floors are easy to clean with high pressure water and detergent. Attention should be paid to having some fall on the concrete runs so the water and mess can drain away quickly and easily when hosed out. Some of the better runs incorporate a septic system to dispose of the waste. Moveable kennels must be relocated every few months to prevent parasite infestation. I would also recommend treating dogs for worms and fleas when the move occurs.

Feeding

The nutrition of working dogs is very important. Many people do not understand this and often I have commented on the lack of condition on people's dogs. I have been told: 'They're working dogs, they should look poor.'

This is a popular misconception. Sure a dog shouldn't be a porker but it should be in good strong muscular condition, much like a top athlete.

This is achieved by a correct diet and a proper worming program. There are a number of top quality prepared feeds on the market today. These feeds usually consist of sorghum, meat meal, soya beans and tallow and include all the necessary vitamins and minerals required to keep your dog in peak physical condition.

Working dogs require a balance of the five major food types: proteins, carbohydrates, fats, vitamins and minerals. I use prepared dry food for my dogs and I find it keeps them in good condition. I don't feed my dogs much meat as I have found a full meat diet reduces their ability to work and makes their condition poorer. Dogs need to be fed only once daily. I feed mine at night because if you feed in the morning the dogs seem to want to lie around and go to sleep with a belly full of food.

Protein is required for growth; pups and lactating bitches require a higher protein intake than normal. Proteins build and maintain cells, however in cases of starvation can be broken down to provide energy. Proteins account for more than 50 per cent of a dog's dry weight.

Carbohydrates include sugar, starch, dextrin, cellulose and glycogen. Carbohydrates provide the main structural and energy storage facilities. When energy is required they are broken down by enzymes.

Fats produce more than twice the energy of carbohydrates. Fats are easily stored in the body for use when carbohydrates are in short supply.

Vitamins and minerals are needed to enhance the animal's metabolism of proteins, carbohydrates and fats; without them the breakdown of foods cannot occur.

12

Diseases of the Working Dog

12

The most common diseases of working dogs are internal parasites — hookworms, roundworms and tapeworms. They are particularly serious in young puppies, causing anaemia, diarrhoea, dehydration and weight loss.

Viral infections, such as parvo, canine distemper and canine hepatitis should be prevented with vaccinations when the risk of infection is high.

Internal Parasites

Hookworms

These are the more dangerous intestinal worms. The larvae attaches itself to the wall of the gut and sucks blood from the intestinal walls. Adult worms also suck the blood. The larvae mature in 2–3 weeks into white adult worms 10–20mm long. Female worms lay eggs which are passed in the dog's droppings. These hatch into larvae within one week to continue the cycle.

Methods of Infection
- *Oral:* The larvae in already infected dog's droppings are swallowed and travel the digestive system to the gut where they attach themselves to the gut wall.
- *Through the skin:* The larvae can burrow through the skin, and migrate through the tissues to reach the gut.
- *Via the uterus:* Larvae in the uterus of a pregnant bitch can migrate to the foetus.
- *Via the mammary gland:* Dormant (sleeping) larvae in the mammary gland become active and pass to the suckling pups in the milk.
- *From rodents:* Rats and mice can become infected. When a dog eats them, dormant larvae in their tissues complete development in the dog.
- *Through inhibited stages in the gut wall:* Larvae can burrow into the gut

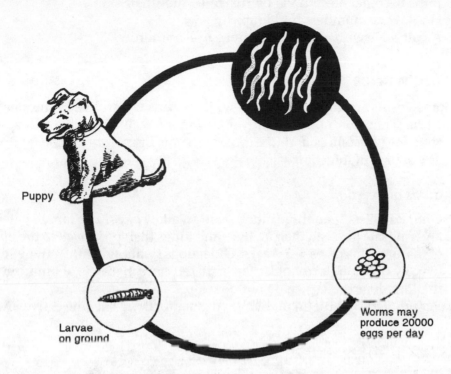

Hookworm life cycle

wall and remain dormant for an indefinite period. They act as a reservoir for re-infection once adult worms are removed by an effective treatment.

Damage Caused by Hookworms

Skin penetration and tissue migration cause inflammation. The stages in the gut produce anaemia, enteritis, diarrhoea, dehydration and, in young pups, death.

Female worms can each lay 20 000 eggs daily; larvae can survive on the ground for months; the development time (from egg to adult female laying eggs) is only 3–4 weeks.

Pups can be infected at the rate of thousands of worms per day, hence the many deaths seen in un-wormed litters kept on dirt runs. Chronic infection impairs immunity in older dogs and severe infection may result.

Hookworms — Points to Remember

- Hookworms suck blood. Dogs die from anaemia.

115

- Pups become infected via uterus, milk, mouth or skin.
- Hookworm numbers build rapidly.
- Adult dogs with chronic infections lose immunity.

Roundworms

Roundworms are large (10–18 cm) white worms which live in the small intestine. They don't permanently attach to the wall. There are two species but one (*Toxocara canis*) is the more common. Eggs from female worms are the source of infection.

Methods of Infection

- *Oral:* Pups less than three months old swallow eggs. The larvae hatch and migrate through the gut, liver and lungs then pass back to the gut to mature. This takes 4–5 weeks. Older dogs swallowing infective eggs usually develop dormant larvae in their tissues. Because of immunity in older dogs the larvae do not develop.
- *Via the uterus:* In pregnant bitches dormant larvae become active and

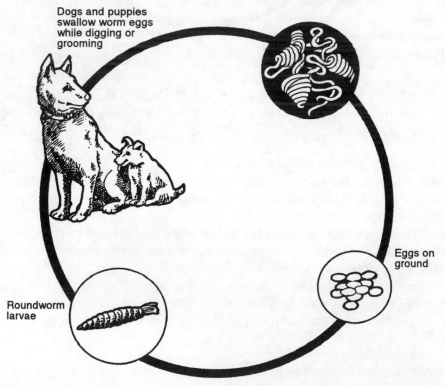

Roundworm life cycle

pass to the foetus in the last fortnight of pregnancy. When the pups reach three weeks of age adult worms are present in their gut. Some larvae remain dormant in the mother and can infect pups in future pregnancies.

- *Via the mammary gland*: This phenomenon seen in the hookworm's life cycle also occurs with roundworms, but is less dangerous.
- *From re-infected bitches*: Because of their decreased immunity during pregnancy, or because they swallow larvae when grooming the pups, lactating bitches have been found to pass infective eggs via their droppings.
- *From rodents*: If dogs eat rats or mice carrying dormant larvae they become infected.

Damage Caused by Roundworms

Larvae cause damage while burrowing through the lungs on their migration to the gut. Vomiting, diarrhoea, 'pot-belly' and colic are common with roundworm infestation of pups.

Roundworms — Points to Remember

- Roundworms produce vast numbers of eggs which survive in the environment for months.
- The bitch is important in infection, contaminating her pups via the uterus, milk and droppings and carrying dormant larvae for the next pregnancy.

Whipworms

Whipworms live in the lower bowel. The worm is 4–7 cm long with a thin neck and stumpy tail which gives it a whip-like appearance. The female worm burrows into the wall of the bowel and produces over 2000 eggs per day. These are passed in the droppings, become infective in three weeks. When eggs are swallowed larvae hatch out and burrow into the small intestine. Soon they migrate to the lower bowel where they mature. The whole process takes 10–15 weeks.

Method of Infection
- *Oral*

Damage Caused by Whipworms
Inflammation of the bowel caused by the burrowing of the worm produces pain and if there are large numbers, diarrhoea.

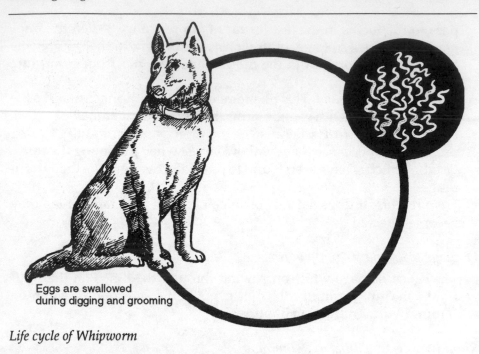

Eggs are swallowed
during digging and grooming

Life cycle of Whipworm

Whipworms — Points to Remember
- Eggs last for years on the ground.
- Adult whipworms live for over a year.
- Dogs with light infection can cause a build-up of contamination in their surroundings over periods of time.

Tapeworms

These are long flat worms, their bodies divided into repeating segments. There are five species of importance, two of which — the common flea tapeworm and the hydatid tapeworm — are of major importance.

Tapeworms have an indirect life cycle, the larval stage must develop in an intermediate host animal.

Common Flea Tapeworm
The worm is found in most dogs, the intermediate host being the flea. When a dog eats an infected flea, the tapeworm develops in the gut. It occurs in large numbers in the dog and may be 50 cm long.

Damage caused by the flea tapeworm.
This worm produces few health problems for the dog. Irritation leads to

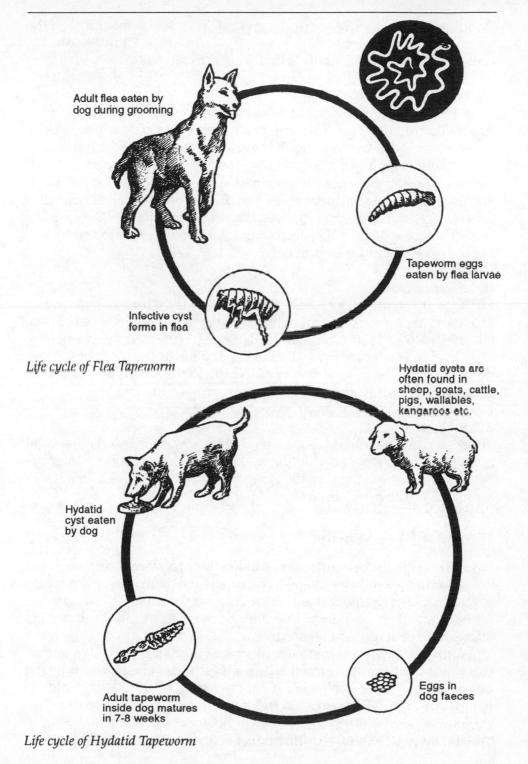

Adult flea eaten by
dog during grooming

Tapeworm eggs
eaten by flea larvae

Infective cyst
forms in flea

Life cycle of Flea Tapeworm

Hydatid cysts are
often found in
sheep, goats, cattle,
pigs, wallables,
kangaroos etc.

Hydatid
cyst eaten
by dog

Adult tapeworm
inside dog matures
in 7-8 weeks

Eggs in
dog faeces

Life cycle of Hydatid Tapeworm

119

skidding of the backside along the ground. This, and appearance of the worm segments around the dog's anus, is unsightly. Children can become infected by accidentally swallowing infected fleas.

Hydatid Tapeworm

Found in areas where sheep and dogs are in contact, this worm is very small (5 mm) but poses a serious risk to man. The cyst which usually develops in sheep from eating the eggs passed by an infected dog, can also occur in man. Its treatment in man requires radical surgery. Cysts in sheep, cattle and pigs mean condemnation of offal/carcases at slaughter and therefore represent an economic loss. Each worm produces vast numbers of eggs, and each cyst in a sheep can provide millions of new worms to a scavenging dog eating sheep carcases. Humans are infected by eating the microscopic tapeworm eggs.

Other Tapeworms

Includes sheep measles (*Taenia ovis*) and bladder worm (*T-hydatigena*). These worms have sheep or rabbits as their common intermediate hosts, but may also be found in goats, cattle or pigs. Infective cysts eaten by dogs lead to tapeworm infestation in the dog's intestine. Whilst the dog does not suffer because of this, cysts in cattle, sheep and pigs are of economic importance as they lead to offal/carcase condemnation.

Zipper Worm

This parasite causes sparganosis, an inflammation of infested muscles in man, when undercooked wild pig is eaten. Pigs are secondary intermediate hosts carrying the juvenile worm in their muscle. Dogs do not suffer severe health problems from this parasite.

Treatment For Worms

Note: While the above points may be indicative of a worm infection, some can also indicate other diseases. The expert in diagnoses of the worm species, and the extent of the problem, is your veterinarian. He/she has the equipment and expertise to examine your dog's motions under the microscope to diagnose the infection.

Worm infections are particularly serious in young pups. It's important to realise that pups may have heavy infections of hookworms by the time they are 14–17 days old, of roundworms by 19–23 days old, of whipworms by 10–12 weeks old and tapeworms by 4–6 weeks old.

Because roundworms and hookworms are continually maturing in the pup's intestine worm treatment is needed more frequently than for adult dogs. Pups should be treated at 2, 4, 8 and 12 weeks old then at 4, 5, and 6 months.

Adults should be treated every three months.

Breeding/lactating bitches should be wormed 10 days before mating, 10 days before whelping, and 2 and 4 weeks after whelping to decrease infection in the puppies. Remember also that worm treatments are not vaccines, they do not protect against re-infection. If the environment of the dog is heavily contaminated with larvae or eggs, then a worm burden can re-occur within 3–4 weeks of an effective treatment.

Intestinal worms — What to Look For

Because of their different behaviour during development and as adults, worm species cause a variety of signs of disease in dogs. In general, however, the main signs of a worm infection include:

- *Anaemia* — paleness of skin and gums. This is due to the blood-sucking activity of hookworms.
- *Diarrhoea* — caused by irritation to the wall of the gut by worms.
- *Dehydration* — seen as a dry skin / coat.
- *Pot-belly appearance* — this occurs commonly in young puppies with a heavy roundworm infection.
- *Weight loss* — can occur even while the dog is showing a ravenous appetite.
- *White segments in droppings/coat/around anus* — occurs with tapeworm, especially the common flea tapeworm.
- *Tail skidding* — dogs with tapeworm infection will often sit or drag their tail round the grass.

Re-infection Following Worm Treatment

Worm treatments are not vaccines, Droncit and Drontal will remove parasites present in the intestine at the time of treatment, but will not protect against re-infection. If the dog's environment is heavily contaminated with eggs or larvae, or in the case of the common flea tapeworm, a high flea population, then a worm burden can re-occur within 2–3 weeks after using Droncit or Drontal.

Because roundworms and hookworms migrate through the dog's body before maturing in the intestine, treatment is needed more frequently in pups and young dogs. Only the parasites present in the intestine will be killed by currently available treatments.

If your dog continues to pass tapeworm segments in the first two weeks after treatment with Droncit or Drontal, these will belong to the zipper worm. Use Droncit at 4 times the usual dose rate to remove this parasite.

The dog's environment can remain a continual source of re-infection. Careful attention to picking up the dog's faeces, combined with regular worm treatment and flea control measures will help reduce the problem.

Generally speaking it takes three months before the intestinal worm burden has built up enough to warrant treatment. However, where many dogthe s are housed together re-infestation is rapid and more frequent worming is necessary.

Zoonoses — The Human Health Risk from Dog Worms

Zoonoses is the name given to diseases that can be transmitted between animals and man. Many people are ignorant of the fact that dog intestinal worm species can be transmitted to man.

Apart from hydatid tapeworm disease, the overall incidence of these zoonoses is relatively low. However to safeguard the health of the family, it is important to take appropriate preventive measures, which include:

- *Good hygiene.* Infants should be kept away from domestic pets until they can practise good hygiene. Children should be taught to avoid long close contact with dogs and always wash their hands and face thoroughly after playing with their pets. Train children not to eat dirt.
- *Treat dogs regularly with an effective worming product* such as Drontal Allwormer.

Hydatid Disease Prevention

Hydatid disease poses a serious health risk for people living in rural and semi-rural areas. Dogs become infected with hydatid tapeworm from eating the internal organs of sheep, cattle, pigs, kangaroos and wallabies. This small tapeworm produces thousands of eggs which are passed in the dog's faeces. These eggs are usually ingested by grazing animals, but if swallowed by people cysts develop in one of the internal organs — liver, lung or brain. Dogs can harbour thousands of the tapeworms in their intestine without ill-effect, and usually have large numbers of tapeworm eggs on their coat which continually contaminate the environment in which they live.

To protect you and your family from hydatid disease it is important to follow these simple rules.

- Dose dogs for hydatid tapeworm every six weeks alternately with Droncit and Drontal.
- Never allow dogs to eat offal (internal organs of sheep).
- Keep dogs under control. They should be tied up, kennelled or shut in an escape-proof yard so they cannot scavenge for dead carcasses.

- Wash your hands after handling dogs, especially before eating or smoking.

Heartworm

This disease is spread by mosquitos from infected dogs. If you travel with your dog to the city or to the coast, your pet should be on heartworm preventative tablets.

Prevention
Before starting on heartworm prevention, your dog should have a simple blood test to make sure it is free from heartworm. Allergic reactions that are occasionally fatal can occur if preventative tablets are given to dogs already infected with heartworm.

Heartguard is a tablet taken once a month. It also comes in chewable form. Dimmitrol is a daily treatment.

Heartworm prevention should begin at 6 weeks of age and must continue for the whole of the dog's life.

Viral Infections In Dogs

Parvo Virus

Parvo virus is a contagious gastroenteritis of dogs. Parvo virus first appeared worldwide in 1978 and is related to a viral enteritis of cats. A slight alteration of the virus allowed it to infect dogs, but cats can't get parvo from dogs.

Spread
Parvo is spread through the droppings of infected dogs and can survive in yards, parks and streets for up to a year. It can also be spread from area to area by anything, shoes, clothes etc., contaminated with infected droppings.

Signs of Parvo
Parvo is usually seen in dogs that are less than 8 months old and have not been fully vaccinated as pups. Dogs are generally depressed and may show vomiting, diarrhoea, often with blood in their motions, and dehydration. This is because the virus attacks cells lining the intestine that absorb fluid into the bloodstream. Parvo can be fatal.

The above signs are a strong indication of parvo but your veterinarian can also do a test to confirm the signs.

Treatment
Treatment involves replacing the fluid that is lost due to vomiting and diarrhoea. An intravenous drip and drugs are the best way to do this. A dog with parvo may be hospitalised for up to five days. Once home it should be fed a specially formulated food to help settle things down.

Prevention
A full vaccination schedule is needed once your dog is well. Remember, parvo lasts in the environment for up to a year.

Canine Distemper

This is a serious disease of dogs and ferrets characterised by fever, discharges from the nose and eyes, vomiting, diarrhoea and pneumonia.

In a number of dogs, nervous signs are also seen, although somewhat later in the course of the disease. These may be muscle spasms, convulsions and progressive paralysis. In dogs the mortality rate can be high. Those surviving may have permanent brain damage.

Canine Hepatitis

This disease is characterised by a loss of appetite, depression, diarrhoea (often with blood), tonsillitis and acute abdominal pain, because of the enlarged liver. The disease may be severe, with death occurring within twenty four to thirty six hours of onset; or mild, the dog showing only signs of loss of appetite and general lethargy.

The severe form is rare in dogs over two years of age.

Kennel Cough (Infectious Tracheobronchitis)

This is a contagious disease of dogs which has more than one cause. Infected dogs have a hacking cough which usually appears after exercise and may persist for several weeks. Among the infectious agents associated with kennel cough are a bacterium, *Bordetella bronchiseptica* and two viruses, *Canine adenovirus type 2* and *Canine parainfluenze* virus. Treatment with antibiotics sometimes speeds recovery.

Tetanus

This disease is rare in dogs and cats. It strikes when penetrating wounds become infected with the tetanus organism. Toxin is produced in the wound and spreads to affect nerves, resulting in muscle stiffness, tremors and 'lock jaw'.

Diagnosis and Treatment of Viral Infections

Your veterinarian should be consulted as the signs of these diseases often vary and complicated laboratory tests may be needed before a diagnosis can be made.

A drug with safe and effective antiviral activity has not been developed. Treatment consists of supportive and symptomatic therapy and often hospitalisation. Intravenous fluids, antiemetics or antidiarrhoeal drugs may be needed. Regretfully treatment is not always successful.

Prevention is Most Important

Since it is difficult, if not impossible, to effectively isolate or quarantine your animal from exposure to these viral diseases, the only effective method of prevention is vaccination. Vaccines stimulate the animal's immune system to produce antibodies against the specific viruses. The antibodies remain in the bloodstream for varying periods of time; it is therefore important that follow up booster vaccinations are given throughout the animal's life.

Response to Vaccinations

Young animals are temporarily protected against many diseases by antibodies received through their mother's colostrum (first milk). These maternal antibodies may also neutralise vaccines. The young animal will respond to vaccination when these maternal antibodies decline to a sufficiently low level. This will occur at any time from 6–16 weeks of age and varies from animal to animal even within the same litter.

In areas of high risk, animals should ideally be vaccinated every two weeks from 6–16 weeks of age. This will ensure that when the animal is able to respond to vaccination, there will be vaccine present to stimulate a response. The period when an animal would be unprotected, by either maternal or vaccine induced immunity is thus minimised.

However, vaccination every two weeks is normally too expensive and impractical to be used. Vaccinations at 6–8, 12–14 and 16–18 weeks is normally satisfactory. However, not every animal is protected until after the third dose.

Fleas

Fleas are a major cause of irritation to dogs. They can also cause nuisance and inconvenience to handlers and their families.

On-dog flea control products are often used with the expectation they will be all that is required to control a flea outbreak. But unfortunately because of the nature of the flea life cycle there is a massive reservoir of eggs, larvae and pupae hundreds of times larger than the number of fleas we find on our dogs. Therefore it is most important to get fleas under control both on your dogs and in the environment.

The dog flea (*Ctenocephalides canis*) and the cat flea (*Ctenocephalides felis*) are both able to live on dogs or cats and are both able to use humans as an intermediary in their life cycle. In Australia the cat flea is the most common.

Both adult males and adult females are parasitic. They feed exclusively on blood, ingesting up to 20 times their own body weight. The blood intake is only partially digested and passes quickly through the fleas short gastro-intestinal tract and is deposited on the coat as a fine excrement.

Dogs that are infested by fleas suffer a range of symptoms. Frequent scratching and biting of the coat is evident. Wounds that may be caused by severe scratching are subject to secondary bacterial invaders.

In some animals there is a severe reaction to the antigenic effects of fleas' saliva and even a small number of bites can set up the condition known as 'flea allergy dermatitis'.

The flea life cycle

Fleas have a life cycle that can be extremely rapid, taking as little as two weeks under ideal conditions.

Eggs: During its lifetime the female lays 300 to 500 eggs in several batches in the host animal's coat or in the environment. Kennels or resting places can become heavily loaded with eggs. Depending on temperature conditions a minimum of three days to three weeks pass before larvae hatch from the eggs.

Larvae: The larvae are about 6mm long, covered with bristles. They are sensitive to light so they tend to live in dark areas, bedding, cracks in timber or in the soil. They feed on organic waste present in their surroundings. There are three larval stages, at the third stage the larvae envelop themselves in a cocoon forming a pupae.

Pupae: These are very difficult to see because soil particles or dust and

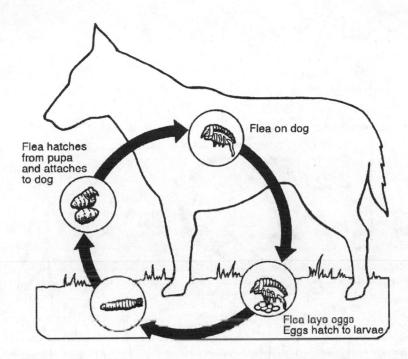

Life cycle of the flea

fibres adhere to the surface providing camouflage. Under warm conditions this pupal stage can take only one or two weeks, but in colder periods it takes longer.

In the absence of animals a large proportion of the pupae remain dormant. Certain environmental stimuli, especially vibration caused by animals, induce the adult flea to emerge from the cocoon.

To control fleas, both the dog and the environment have to be treated. Repositioning the kennel if possible, is ideal. Treatments are commercially available for both the dog and the environment and repeat treatments are necessary as most chemicals do not destroy the eggs.

Mites

Mites are microscopic parasites, there are three main types that effect dogs — itch mite (fox mange, mange mite) (*Sarcoptes scabici*), ear mite (*Otodectes cynotis*) and follicle mite (*Demodex canis*). They feed on skin debris; the itch mite actually burrow into the skin where they live and breed. When their eggs hatch the larvae either move to fresh skin or stay

127

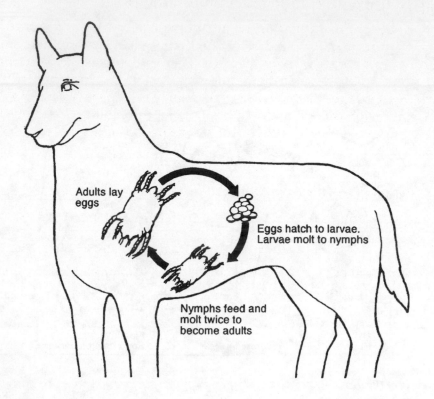

Life cycle of the mite

in the same area and burrow further along the skin. Mites cannot survive for long when away from body heat and moisture as they need both to survive. Mites are spread by body contact, such as nursing bitch to pup. If you suspect your dog has a mite infestation contact your vet immediately for an appropriate course of action.

Follicle mite (*Dermodex canus*)

These mites are the cause of dernodectic mange, the second most common form of mange found on dogs. It occurs mainly on dogs in poor condition or on young dogs. If left untreated it will be fatal. Signs of demodectic mange are loss of hair and wrinkled skin, abscesses will appear as secondary infection as the mite becomes more prolific. If left untreated skin lesions will become more widespread and eventually cover the dog. If you see any of these symptoms contact your vet.

Mange mite, itch mite (sarcoptic mange)

Mange mite or fox mange burrow into the skin of the dog to feed and lay eggs. They are usually found in the head, ears, neck and shoulders, however they can spread all over the body. These mites cause severe itching and cause sores to develop on the body. As the skin becomes more tender it cracks and leaves deep wounds that usually become infected. The dog will continually scratch and bite itself, its hair will fall out and its skin will become wrinkled. Once again, if you suspect your dog has fox mange consult your vet.

Ear mite (*Otodectes cynotis*)

Ear Mite as the name suggests appears on the outside of the ears, causing severe irritation. As the infestation increases the ears will droop and show a discharge. Treatment is necessary immediately or the infection could spread to the inner ear. Secondary infection is also probable. Veterinary assistance is required to treat your dog.

Conclusion

As I mentioned at the start of the book, training working dogs can be both frustrating and exceptionally rewarding. The methods mentioned in the text are all tried and proven and I have enjoyed great success employing them. They are, however, by no means the only methods used. Experience is a great teacher and as you work with dogs undoubtedly will discover many variations.

This text has not attempted to include any special skills training for trial work as this level of expertise is only learnt over time and with experience and an understanding of your particular dog's temperament and ability. The section about diseases is a guide and is included for general knowledge only. If your dog is sick seek veterinary assistance. As a general guide however the material included will provide any handler the tools necessary to train more than useful yard and paddock dogs. Good luck and I wish you all every success.

Glossary

backing	the dog's ability to run along the animals' backs
bow hocks	hocks bowed outward (opposite to cow hocked)
broken in	fully trained
brush	bushy hair on the underside of the tail
casting	to move in a wide arc around the mob
close casting	to cast too close to the mob
small casting	cast while handler is in close proximity to the mob
wide casting	to run in a wide arc to a mob at a distance
short caster	a dog that won't complete its cast, it stops short of a full cast
draw	the dog's ability to bring the stock to the handler
driving	the dog's ability to move stock away from the handler
eye	the dog's ability to hold sheep still by looking at them
too much eye	the dog becomes intent on staring at the sheep and tends not to respond to commands
hard eyed	the dog's stare is so intent on the sheep it appears to be mesmerised and won't respond to commands
sticky eyed	hard eyed
strong eyed	hard eyed
loose eyed	has no ability to hold the sheep with eye
feather	the longer hair on the backs of the dog's legs
head	the direction the stock are moving
heading dog	a dog that instinctively runs to the head of the mob to block its progress
hocks	the lower bone of the back leg
lead	the sheep leading the mob or at the head of the mob
lead dog	a dog that will slow down but not stop the progress of a mob, used for droving

leather	the skin on the dog's ear
lift	initial movement of the mob of sheep towards the handler
lift and draw	the initial movement of the mob (lift) and the continued movement to the handler
look	a command that instructs the dog to look for stock it has missed when mustering
pasterns	bones between the forearm and the foot
pencil	to keep score for the judge in dog trials
race	long narrow pen
running the race	run along the race outside the fence
ring	run right around the mob
root	where the tail grows from the body
set-on	the angle on the tail
stifles	stifle joint, the joint between the upper and second thigh or gaskin in the back leg
shepherding	the dog's ability to keep the stock in a mob
stick up	the dog's stare is so intent on the sheep it appears to be mesmerised and won't respond to commands

Index